p-Adic Valued Distributions in Mathematical Physics

Mathematics and Its Applications

Volume 309

p-Adic Valued Distributions in Mathematical Physics

by

Andrei Khrennikov
Moscow Institute of Electronic Engineering,
Zelenograd (Moscow), Russia

KLUWER ACADEMIC PUBLISHERS
DORDRECHT / BOSTON / LONDON

A C.I.P. Catalogue record for this book is available from the Library of Congress.

ISBN 978-90-481-4476-1

Published by Kluwer Academic Publishers,
P.O. Box 17, 3300 AA Dordrecht, The Netherlands.

Kluwer Academic Publishers incorporates
the publishing programmes of
D. Reidel, Martinus Nijhoff, Dr W. Junk and MTP Press.

Sold and distributed in the U.S.A. and Canada
by Kluwer Academic Publishers,
101 Philip Drive, Norwell, MA 02061, U.S.A.

In all other countries, sold and distributed
by Kluwer Academic Publishers Group,
P.O. Box 322, 3300 AH Dordrecht, The Netherlands.

Printed on acid-free paper

*This book is devoted to
professor V. S. Vladimirov*

Table of Contents

Introduction

Numbers ..., natural, rational, real, complex, p-adic ... . What do you know about p-adic numbers? Probably, you have never used any p-adic (nonrational) number before now. I was in the same situation few years ago. p-adic numbers were considered as an exotic part of pure mathematics without any application. I have also used only real and complex numbers in my investigations in functional analysis and its applications to the quantum field theory and I was sure that these number fields can be a basis of every physical model generated by nature. But recently new models of the quantum physics were proposed on the basis of p-adic numbers field $\mathbb{Q}_p$.

What are p-adic numbers, p-adic analysis, p-adic physics, p-adic probability? p-adic numbers were introduced by K. Hensel (1904) in connection with problems of the pure theory of numbers. The construction of $\mathbb{Q}_p$ is very similar to the construction of (p is a fixed prime number, $p = 2, 3, 5, \ldots, 127, \ldots$). Both these number fields are completions of the field of rational numbers $\mathbb{Q}$. But another valuation $|\cdot|_p$ is introduced on $\mathbb{Q}$ instead of the usual real valuation $|\cdot|$. We get an infinite sequence of nonisomorphic completions of $\mathbb{Q}$: $\mathbb{Q}_2, \mathbb{Q}_3, \ldots, \mathbb{Q}_{127}, \ldots, \mathbb{R} \equiv \mathbb{Q}_\infty$. These fields are the only possibilities to complete $\mathbb{Q}$ according to the famous theorem of Ostrowsky.

In 1968 two pure mathematicians, A.F. Monna and F. van der Blij, proposed to apply p-adic numbers to physics. It was not a rigorous physical investigation. The only motivation was that every "good" mathematical construction must correspond to some physics. But this point of view does make sense. Then in 1972 E. Beltrametti and G.

Cassinelli investigated a model of p-adic valued quantum mechanics from the positions of quantum logic. Their result was negative (according to standard formalism) and they did not continue research in this direction. And V.S. Vladimirov and I.V. Volovich applied p-adics to superfield theory in 1984. All these investigations were independent and separate as the authors worked in very different directions.

Much interest in p-adic physics was shown in the connection with string theory. The paper of I.V. Volovich (1987) generated a series of papers on the p-adic string theory. The main idea of Volovich was to exchange the usual real space-time $\mathbb{R}^4$ to a p-adic space-time $\mathbb{Q}_p^4$ and to study a string in this space-time. A number of models of p-adic strings were investigated by P.G.O. Freund, M. Olson, E. Witten, P.H. Frampton, G. Parisi and many other authors. It was a great program of the p-adic string (p-adics were used in a different way in these models). There were a number of very interesting results from the mathematical point of view. But a complicated problem of physical interpretation of the proposed formalisms was not resolved until this time. For this reason, simpler models such as a p-adic quantum mechanics and field theory were studied in 1988–1990.

What is the main difference between real and p-adic space-time? It is the Archimedean axiom. According to this axiom any given large segment on a straight line can be surpassed by successive addition of small segments along the same line. This axiom is valid in $\mathbb{R}$ and it is not valid in $\mathbb{Q}_p$. However, it is a physical axiom which concerns the process of measurement. To exchange a number field $\mathbb{R}$ to $\mathbb{Q}_p$ is the same as to exchange axiomatics of the measurement in quantum physics. This point of view was discussed by I. V. Volovich in 1987. It was an old idea that gravitational measurement cannot be localized in a sub-Plancian domain (at incredibly small distances of the order 10^{-33}). Roughly speaking if we try to measure a segment l of an order 10^{-33}, the only possibility is to use the electromagnetic (or other kind) waves to compare a length of a wave with l. This is why, for example, some kind of electromagnetic emmition must be used. But according to the principle of the equivalence of the pictures of waves and particles in the quantum field theory, it is the same to send particles (photons) to some fixed volume V of the space which contains l. Thus some quantity of an energy is sent to V. Now according to the principle of the equivalence of energy and mass, mass in the volume V is changed. And according to the principle of the equivalence of mass and geometry, this measurement would change geometry in V and hence, change l. Hence, we should never measure l with the aid of our unit of a measurement. A new idea was to use the p-adic space-time to describe this nonlocalization.

This problem will be discussed carefully in the book. I will also present another idea: *p-adic numbers may be useful to measure infinite-large cosmic distances (or infinite-large energy levels, number of particles ...).* However, this idea is not in strong contradiction with the idea to use p-adic numbers to describe a sub-Planckian domain. We shall present an infinite scale of observers and a distance which is an infinite-small from the point of view of one observer can be infinite-large for another observer. These problems are much more philosophical than mathematical and they are not the main problems studied in the book.

Attention is paid mainly to mathematical problems generated by the non-Archimedean physics. An extended mathematical apparatus is developed for the physics involved. A theory of non-Archimedean valued distributions is the basis of this apparatus. A theory of evolutional differential equations, a theory of integration and a harmonic analysis are proposed on the distribution basis. A theory of non-Archimedean pseudodifferential operators (PDO) is developed within this theory of distributions. On the basis of PDO calculus we realize the operator quantization for a quantum mechanics with non-Archimedean valued wave functions. The most important differential equations of the non-Archimedean valued mathematical physics such as heat, Schrödinger's, Heisenberg's and Liouville's equations are investigated in spaces of test functions and distributions (generalized functions).

We are also interested in non-Archimedean fermions and a corresponding mathematical apparatus. A non-Archimadean variant of superanalysis is developed on the basis of infinitely-generated Banach algebras.

As the quantum field theory is the theory allows infinite degree of freedom, we need to propose an infinite-dimensional non-Archimedean analysis if we wish to study a quantum field theory with non-Archimedean valued fields. Such analysis is also presented in the book. The main results are in a theory of infinite-dimensional distributions with applications to the problem of the functional integration. The quantization of a Bosonic non-Archimedean valued field is carried out in the functional integral formalism.

The main problem in the physical understanding of the presented mathematical formalism is a problem of the probability interpretation of non-Archimedean valued wave functions. Coefficients in an eigenfunction expansion belong to the quadratic extensions of $\mathbb{Q}_p$. To propose any probability sense to these coefficients we need to develop a theory of probability where probabilities of events can belong not only to the segment $[0, 1]$ of the field $\mathbb{R}$ but also to some subset of $\mathbb{Q}_p$. It would be impossible to realize such theory of probability within Kolmogorov's accepted axiomatics based on the real valued measure theory. New non-Kolmogorov theory of probability was created. The foundation of this theory is a general principle of the statistical stabilization extending the standard one. According to this principle, a statistical stabilization of relative frequencies can be studied not only with respect to the real topology (metric) on the field of rational numbers $\mathbb{Q}$ but also with respect to every another topology on $\mathbb{Q}$. The real metric on $\mathbb{Q}$ is only one of possible metrics of the statistical stabilization. New statistical models are presented. Relative frequencies oscillate between 0 and 1 with respect to the real metric and stabilize with respect to one of the p-adic metrics. Such models cannot be a subject of the usual theory of probability but they may be studied with the aid of the non-Kolmogorov theory of probability. Applications of the non-Kolmogorov theory can be considered not only in quantum physics , but in many other sciences, especially in biology, and possibly in sociology. The general principle of the statistical stabilization of relative frequencies is a new possibility to find a statistical information in the chaotic (from the real point of view) sequences of frequencies.

To finish the introduction we wish to point out that a number of widely different physical and mathematical theories and formalisms is mixed in this book.

Firstly, there is non-Archimedean physics. Here I was under great influence of Vladimirov and Volovich. Their lectures on a p-adic mathematical physics at Tashkent, 1989, attracted my attention to p-adics. But my approach to non-Archimedean physics differs from their idea to quantize with the aid of complex valued functions of p-adic variables (the same idea was independently generated by Freund, Olson, Witten). This book is devoted to the theories with non-Archimedean valued wave functions.

Secondly, there is a non-Archimedean analysis. Our results are strongly connected with formalisms of non-Archimedean valued harmonic analysis (B. de Mathan, Y. Amice, A. Escassut, W. Schikhov, C. Woodcock). The old results of N. Bourbaki and J. Serre were also very useful in our non-Archimedean investigations.

The next is infinite-dimensional analysis and theory of distributions on infinite-dimensional spaces. Here I had a long collaboration with the scientific school of O. G. Smolaynov.

Further, theory of functional integration used here is based on the definition without limiting procedure with the aid of Parseval's equality on infinite-dimensional spaces. Different variants of this definition were introduced by S. Albeverio and R. Hoegh-Khrohn, C. de Witt-Morette, A. V. Uglanov.

The non-Kolmogorov theory of probability was created under the influence of the frequency theory of probability of R.von Mises (1919). The measure-theoretical approach to p-adic valued theory of probability is based on Monna-Springer integral with respect to p-adic valued measures.

I would like to thank V.S. Vladimirov, O.G. Smolaynov, I.V. Volovich, B. de Witt, C. de Witt-Morette, Yu.V. Prohorov, V.M. Maximov, A.N. Shirayev, S. Albeverio, B. Dragovic, R. Cianci, P. Frampton, L. Biedenharn, E. Beltrametti, G. Cassinelli, G. Mackey, W. Schikhov, M. Endo, C. Piron, P. Mittelstaedt, T. Hida, M. Hitsuda, G. Parisi, Z. Huang for the fruitful discussions.

I would also like to thank my wife Olga and daughter Polina for their help.

This book was written with the support of GNFM of CNR, Italy, of the research program "Metodi geometrici in Relativita e teoria dei campi" of MURST (1992, 93), and of the Russian Fundamental Foundation with the project n.93-011-16114, the National Natural Science Foundation of China (1992, 93) and Alexander von Humboldt Foundation (1994, 95).

Mendeleevo-Genova-Beijing-Bochum, 1992-1994.

Chapter I

First Steps to Non-Archimedean

There are introduced the most important notions of non-Archimedean (and in particular, p-adic) mathematics and briefly discussed main results which will be used in the next chapters. There is no any possibility for us to present non-Archimedean mathematics (and in particular, non-Archimedean field theory) step by step in rigorous way proving all results which can be useful for us. The only proofs which will be used in further considerations are presented. This non-Archimedean introduction is very brief. The reader who will wish to study this subject more will be able to continue, for example, with the aid of books of Mahler [97] and Schikhov [106]. Mahler and Schikhov proposed different approaches. Mahler was interested very much in non-Archimedean numbers as *numbers*: digits, expansions, algebraic operations, algorithms of these operations. Schikhov's book is more useful to study non-Archimedean *analysis*. Probably, these books is the simplest way for a nonspecialist to begin to work in non-Archimedean.

1. Non-Archimedean Number Fields

Everywhere in what follows K denotes a complete non-Archimedean field with a nontrivial valuation $|\cdot|_K$. The case $\operatorname{Char} K = 0$ is considered.

Let F be a field. Recall that an *valuation* (in some books *norm*)is a mapping $|\cdot|: F \to \mathbb{R}_+$ satisfying the following conditions:

$$|x|_F = 0 \Longleftrightarrow x = 0, \tag{1.1}$$

$$|xy|_F = |x|_F |y|_F, \tag{1.2}$$

$$|x+y|_F \le |x|_F + |y|_F \tag{1.3}$$

The last inequality is a well known triangle axiom. A valuation is said to be non-Archimedean if the strong triangle axiom is valid, i.e.

$$|x+y|_F \le \max(|x|_F, |y|_F). \tag{1.4}$$

A field F with a non-Archimedean valuation is said to be a non-Archimedean field. An absolute value is non-Archimedean if and only if $|n|_F \le 1$ for all elements n from the ring generated in the field F by its unity element. It follows from the strong triangle inequality (1.4).

Thus, if l and L are two nonzero quantities belonging to K such that $|l|_K < |L|_K$, then it is impossible to find such natural number n that $|nl|_K > |L|_K$. On the other hand if l and L are two quantities belonging to $\mathbb{R}$, there is no problem to find a natural number n such that $|nl| > |L|$, where $|\cdot| \equiv |\cdot|_{\mathbb{R}}$ is the standard valuation on $\mathbb{R}$. This is why we can consider the strong triangle inequality (1.4) as one of the possible mathematical realizations of the non-Archimedean axiom of the measurement theory (see the introduction).

Remark 1.1. As $\operatorname{Char} K = 0$, we can realize the set of natural numbers as the subset of K. The same is valid for the field of rational numbers $\mathbb{Q}$. Everywhere below $\mathbb{Q}$ will be realized as the subfield of K.

A trivial absolute value on F is $|\cdot|_F$ such that $|x|_F = 1$ for all $x \neq 0$.

The absolute value of the field F is a homomorphism of the multiplicative group F^* into a multiplicative group $\mathbb{R}^*$.

We denote by Γ the image of K with respect to this homomorphism, Γ is a subgroup in $\mathbb{R}^*$. For the scalar $R \in \Gamma$, we denote by a_R any element of the field K such that $|a_R|_K = R$.

Every valuation generates a metric. A completeness of K is the completeness with respect to this metric, $\rho_K(x,y) = |x-y|_K$. As the valuation on K is a non-Archimedean one, the corresponding metric has many unusual properties which will be discussed in the next section.

The *field of p-adic numbers* $\mathbb{Q}_p$ can serve as one of the most important examples of non-Archimedean fields. Just as the field of real numbers $\mathbb{R}$, this field is defined as a completion of the field of rational numbers $\mathbb{Q}$, but instead of the standard valuation, we use a new valuation $|\cdot|_p$ called a p-adic valuation. Here p is a fixed prime number, $p = 2, 3, \ldots$, and every p has its own field $\mathbb{Q}_p$ here. It is sufficient to define $|\cdot|_p$ on natural numbers. We factorize a natural n into prime factors: $n = 2^{\nu_2} 3^{\nu_3} \ldots p^{\nu_p} \ldots$, and then $|n|_p = p^{-\nu_p}$ (the minus plays an important role to prove (1.4)). Moreover, by definition, $|0|_p = 0$. If $x = n/m$, where n and m are natural numbers, then $|x|_p = |n|_p/|m|_p$ by the definition.

For instance, $|3|_2 = 1$, $|4|_2 = 2^{-2}$, $|3/2|_2 = 2$ but $|3|_3 = 3^{-1}$, $|4|_3 = 1$, $|3/2|_3 = 3^{-1}$. In particular, if a rational number has the form $x = n/m$ where n and m are integers and p does not divide n and m then $|x|_p = 1$. Thus the p-adic distance between the points of $\mathbb{Q}$ essentially differs from the real distance.

Recall that the valuations $|\cdot|_\alpha$ and $|\cdot|_\beta$ are said to be equivalent, if $|x|_\alpha = |x|_\beta^c$, $c \in \mathbb{R}$, $c > 0$.

Theorem 1.1 (Ostrovsky). *Any nontrivial valuation on the field* $\mathbb{Q}$ *is equivalent either to the real valuation* $|\cdot|$ *or to one of the p-adic valuations* $|\cdot|_p$.

In the non-Archimedean case the function $1/|n!|_K$ increases. We will later need the estimate of this increase. Note the estimate of the field $\mathbb{Q}_p$:

$$(1/np)p^{n/(p-1)} \leq \frac{1}{|n!|_p} \leq p^{(n-1)/(p-1)}$$

Using Ostrovsky's theorem, we find that $|x|_K = |x|_p^l, p = p(K), l = l(K) > 0$, and the exponential estimate

$$a^n \leq \frac{1}{|n!|_K} \leq b^n, \tag{1.5}$$

where $a = a(p, l)$ and $b = b(p, l)$, is valid for any K. We shall use the asymptotics $\frac{1}{|n!|_K} \asymp p^{nl/(p-1)}$ for large n.

The strong triangle axiom implies the simple but very useful

Theorem 1.2. (A dream of a bad student) *The series* $\sum_{n=1}^{\infty} a_n$, $a_n \in K$ *converges in* K *if and only if* $a_n \to 0$, $n \to \infty$.

To prove this result we need only to use the Cauchy theorem in complete metric spaces.

For example, the series

$$1 + p + p^2 + ... + p^n + ...$$

converges in $\mathbb{Q}_p$ as $|p^n|_p = p^{-n} \to 0, n \to \infty$. The sum is computed by usual rule as the limit of finite sums, $S = 1/(1-p)$.

The following proposition is also a consequence of the strong triangle inequality. It will be very useful in some estimates and computations.

Proposition 1.1. *Let* $x, y \in K$. *Then the inequality* $|x|_K \neq |y|_K$ *implies* $|x + y|_K = \max(|x|_K, |y|_K)$.

2. Ultrametrics

We set $\rho_K(x, y) = |x - y|_K$. There is no problem to check up that it is a metric. But using the strong triangle inequality for the valuation, we get that $\rho_K(x, y) \leq \max[\rho_K(x, z), \rho_K(z, y)]$, $x, y, z \in K$.

It is interesting to consider general metric spaces X where instead of the standard triangle inequality

$$\rho(x, y) \leq \rho(x, z) + \rho(z, y), \quad x, y, z \in X,$$

the strong triangle inequality

$$\rho(x, y) \leq \max[\rho(x, z), \rho(z, y)], \quad x, y, z \in X,$$

is valid. Such metric is called ultrametric and such metric spaces are called ultrametric spaces. The strong triangle inequality can be stated geometrically: *Each side of a triangle is at most as long as the longest one of the two other sides.*

Let us discuss main properties of ultrametric space X. We set $U_r(a) = \{x \in X : \rho(x-a) \leq r\}$ and $U_r^-(a) = \{x \in X : \rho(x-a) < r\}, r \in \mathbb{R}_+, a \in X$. These are balls of radius r with the center in the point a. Our standard intuition tells us to said that $U_r(a)$ is a closed ball and $U_r^-(a)$ is an open ball. But it is not so in ultrametric spaces.

Proposition 2.1. *Each ball in X is both open and closed. Each point of a ball may serve as a centre. A ball may have infinitely many radii.*

Proof. Of course, $U_r^-(a)$ is open and $U_r(a)$ is closed. Let us prove, for example, that $U_r(a)$ is open. Let $b \in U_r(a)$, we show that $U_r(b) \subset U_r(a)$ (it is much more then we need to prove that $U_r(a)$ is open). Using strong triangle inequality, we get that if $x \in B_r(b)$, then $\rho(x,a) \leq \max[\rho(x,b), \rho(b,a)]$, so $x \in U_r(a)$, i.e. $U_r(b) \subset U_r(a)$. Now there is also no problem (using the symmetry) to prove that $U_r(b) = U_r(a)$ and, hence, every point b of $U_r(a)$ is its center.

Sets which are open and closed in the same time will play a large role in our further investigations. We use the word *clopen* as the abbreviation for closed and open.

Proposition 2.2. *Let U and V be two balls in X. Then there are only two possibilities:*

1) balls are ordered by inclusion (i.e. $U \subset V$ or $V \subset U$);

2) balls are disjoint.

Proof. If none of these statements were true, we could find elements $a \in U \cap V, x \in U \setminus V, y \in V \setminus U$. Then a would be a center of U and V and

$\rho(y,a) > \rho(x,a)$, since $x \in U$ and $y \notin U$;

$\rho(x,a) > \rho(y,a)$, since $y \in V$ and $x \notin V$,

which is a contradiction.

Proposition 2.3. *Let G be nonempty open subset of X. Then there is a partition of G into balls. More specifically, given $r_1 > r_2 > \ldots > r_n > \ldots > 0$, G can be covered by disjoint balls of the form $U_{r_n}(a_n), a_n \in X$.*

It suffices to use the previous proposition to prove this one.

Recall that a topological space Y is *totally disconnected* if the only subsets of Y that are connected are the empty set and singleton sets

$\{a\}$ (a set O is connected iff $O = A \cup B, A \cap B = \emptyset, A, B$ clopen implies $A = \emptyset, B = \emptyset$).

Proposition 2.4. *An ultrametric space X is totally disconnected.*

Proof. For each $a \in X$ the connected component T of a is contained in every clopen neighbourhood of a. But there exists a basis of clopen neighbourhoods. Hence $T = \{a\}$.

Recall that topological space Y is *zero-dimensional* if for every $a \in Y$ and every neighbourhood U of a there is a clopen subset V such that $a \in V \subset U$.

There is no problem to see that *every ultrametric space X is zero-dimensional.*

A function $f : X \to K$ is continuous if for each $a \in X$ and $\epsilon > 0$ there exists $\delta > 0$ such that $x \in X, \rho(x,a) < \delta$ implies $|f(x) - f(a)|_K < \epsilon$. We shall use very often the fact that a set O is clopen if and only if its (K-valued) characteristic function $\phi_O(x)$, defined by $\phi_O(x) = 1$ if $x \in O$ and $\phi_O(x) = 0$ if $x \notin O$, is continuous.

The symbol $S_r(a)$ denotes the sphere $\{x \in X : \rho(x,a) = r\}$ of radius $r \in \mathbb{R}_+$ with the centre at a. There is also a large deviation and the sphere $S_r(a)$ is not a boundary of $U_r(a)$ or $U_r^-(a)$. It can be possible to prove that the boundary of a ball is empty (see Proposition 2.1).

3. Fields of p-adic Numbers

We begin our introduction to properties of $\mathbb{Q}_p$ to study the problem of an expansion of p-adic numbers into series with respect to degrees of p. Every positive real number a can be written as a decimal fraction

$$a = \ldots + \frac{a_{-k}}{10^k} + \ldots + \frac{a_{-1}}{10} + a_0 + a_1 10 + \ldots + a_n 10^n = a_n \ldots a_0, a_{-1} \ldots a_{-k} \ldots, \tag{3.1}$$

where the *digits* a_j assume only ten possible values: $0, 1, \ldots, 9$. The representation is unique with the one exception when after finitely many terms in decimal expansion, all a_k are equal to 9. Thus, for example,

$$9/10 + 9/10^2 + 9/10^3 + ... = 1 = 1 + 0/10 + 0/10^2 + 0/10^3 + ...,$$

and we have two representations to the base 10 for the same number 1. The decimal expansion can be generalized. If $m > 1$ is any fixed integer, then every positive real number a can be written as

$$a = \ldots + \frac{a_{-k}}{m^k} + \ldots + \frac{a_{-1}}{m} + a_0 + a_1 m + \ldots + a_n m^n = a_n ... a_0, a_{-1} ... a_{-k} ..., \tag{3.2}$$

where the digits a_k assume now the values: $0, 1, ..., m-1$. The representation is unique except again in the case when all a_k for sufficiently large k are equal to $m-1$. Then there is a second representation where from one place on ward, all digits are equal to 0. Thus

$$\sum_{k=f}^{\infty}(m-1)m^{-k} = m^{-f+1} + \sum_{k=f}^{\infty} 0m^{-k}.$$

Positive integers can be written to the basis m as finite sums

$$a = a_0 + a_1 m + \ldots + a_n m^n \tag{3.3}$$

The more general sums

$$a = \frac{a_{-k}}{m^k} + \ldots + \frac{a_{-1}}{m} + a_0 + a_1 m + \ldots + a_n m^n \tag{3.4}$$

represent quotients of positive integers by the sth power of m and are rational. It is well known that a real number a is *rational iff its m-expansion fraction is periodical.*

The situation is sufficiently similar in p-adic case (for a moment we restrict our considerations only to the case of the prime base $m = p$, a general case will be studied later). Every p-adic number a can be represented as a fraction:

$$a = \frac{a_{-f}}{p^f} + \ldots + \frac{a_{-1}}{p} + a_0 + a_1 p + \ldots + a_n p^n + \ldots =$$

$$a_{-f} \ldots a_{-1}, a_0 a_1 \ldots a_n \ldots, \tag{3.5}$$

where $a_j = 0, 1, \ldots, p-1$. This representation is similar to the expansion of a real number, into a fraction but, as distinct from the real case, is infinite in the direction of increase of the powers of p and not in the direction of their decrease.

We prefer to write p-adic expansion as infinite in right direction as the real (3.1) and (3.2). It is not a great problem that there is a difference in order of writing for natural numbers or quotients (3.4) in the real way (3.2) for $m = p$ and in the p-adic way (3.5) (a number (3.3) $= a_n \ldots a_0$ by means of (3.2) and (3.3) $= 0, a_0 \ldots a_n$ by means of (3.5)).

The idea of the proof is clear. First of all every series (3.5) converges in $\mathbb{Q}_p$ as $|p^n|_p = p^{-n} \to 0, n \to \infty$. This expansion is unique as the valuation of zero is zero.

To show the inverse we use that every p-adic number a is a limit of a sequence of rational numbers $\{u_i\}$. If $|a|_p = p^{-f}$ then also $\lim_{i\to\infty} |u_i|_p = p^{-f}$. But since the powers p^{-f} have no limit point other then 0 and ∞, there can be only finitely many terms in $\{u_i\}$ with $|u_i|_p \neq p^{-f}$. These terms may be omitted from $\{u_i\}$ without changing the limit, and hence we may assume that already $|u_i|_p = p^{-f}$ for all i. Then all u_i have the form $u_i = p^f \frac{r_i}{s_i}$, where r_i and $s_i > 0$ are integers with $(r_i, s_i) = 1$ (no common divisors), p does not divide r_i, s_i.

Then it is not a hard problem to prove that each rational number c of such type can be expanded in the form

$$c = c_f p^f + c_{f+1} p^{f+1} + \ldots + c_{N-1} p^{N-1} + p^N (t_N / q_N), \tag{3.6}$$

where the coefficients c_k are digits and $(p, q_N) = 1$. The integer N can be chosen arbitrary large.

To conclude the proof it suffices to show the fundamental property of the sequence $\{u_i\}$ implies that for large i the digits in the expansion (3.6) of u_i are independent of i (and then $N \to \infty$).

The simplest way to realize p-adic computations is to use the canonical expansions of these numbers. The division is realized as the inverse operation to the multiplication. If you wish to get $x = a/b$, then you must expand a, b and x into (3.5) (the last one with undefined coefficients) and resolve step by step an infinite system of equations generated by the equation $xb = a$. We can note that on computer these realizations are much more rapid then the real one.

We shall use canonical expansion of p-adic numbers as the basis of the statistical simulation over fields of p-adic number.

As in real case a *p-adic number is rational iff its canonical expansion is periodical.* It will be very interesting in our physical investigations to get an answer to the question on the rationality of p-adic results. It will be possible to consider these p-adic results as exact physical values. But this problem is a hard problem of the number theory already in the simplest cases. For instance, in view of the estimate (1.5), the series

$$1 + 2! + \ldots + n! + \ldots$$

converges in every $\mathbb{Q}_p$. But there is no results on the exact value of this sum. We do not know is this sum a rational number? There were attempts for a computer investigations to find a periodicity in canonical expansions (in particular, my students have been trying). No periodicity for any p was found.

We note that the convergence in $\mathbb{Q}_p$ is equivalent to a stabilization of digits in the canonical expansion (3.5). This fact will be used in p-adic valued theory of probability to realize a statistical stabilization in $\mathbb{Q}_p$.

The unit ball $U_1(0)$ will play an important role in further considerations. $\mathbb{Z}_p$ is the standard symbol using for $U_1(0)$. In view of strong triangle inequality, $\mathbb{Z}_p$ is an additive subgroup of $\mathbb{Q}_p$. Moreover it is a ring. This ring is called *a ring of p-adic integers.* The motivation is the following. If a natural number is expanded in the canonical way (3.5)=(3.3), then there is no negative degrees of p in this expansion. The same is valid for elements of $\mathbb{Z}_p$.

We shall not write p-adic integers canonical expansion beginning from the zero: $a = 0, a_0a_1 \ldots$. It will be more convenient to write $a = a_0a_1\ldots$ in the cases when there is no problem to understand that we consider only $\mathbb{Z}_p$.

Remark 3.1. The canonical expansion of the negative integer may be an infinite series. For example, let $p = 2$ then

$$-1 = 1/(1-2) = 1 + 2 + 2^2 + ... = 111...$$

Ring $\mathbb{Z}_p$ contain also rational fractional numbers. For example,

$$1/3 = 1 + 2 + 2^3 + 2^5 + ... = 110101...$$

in $\mathbb{Q}_2$.

Let A be a subset in K and $b \in K$. The symbol bA denotes a set $\{x \in K : x = ba, a \in A\}$.

There is no problem to see that $p^n\mathbb{Z}_p = U_{p^{-n}}(0)$. These sets will be very useful in our considerations.

As a simple consequence of the previous section we have that the topology of $\mathbb{Q}_p$ (as every non-Archimedean field K) is totally disconnected and zero-dimensional. By the definition $\mathbb{Q}_p$ is complete and separable.

Proposition 3.1. *The ring of p-adic integers $\mathbb{Z}_p$ is compact*

Proof. Let

$$\{x^k = \sum_{n=0}^{\infty} a_n p^n = a_0^k ... a_n^k ...\}_{k=1}^{\infty} \tag{3.7}$$

be a sequence in $\mathbb{Z}_p$. We show that it has a convergent subsequence. Since there are only finitely many possibilities for a_0^k (namely $0, 1, .., p-1$) we can find $b_0 \in \{0, 1, .., p-1\}$ and subsequence of $\{x^k\}$ such that the first digit in (3.7) is equal to b_0 for the elements of this subsequence. We shall realize the same procedure for the second digit in subsequence and so on. To conclude the prove we shall use the standard diagonal procedure.

Proposition 3.2. *The set of natural numbers is dense in the ring of the p-adic integers.*

The canonical expansion is used to prove this proposition.

Corollary 3.1. *$\mathbb{Q}_p$ is locally-compact*

To analyse a measurement procedure we shall use also generalizations of p-adic numbers. These are the so called *m-adic numbers.* There m is any fixed natural number, $m > 1$. The way to define m-adic numbers is the same to define p-adic numbers. An m-adic pseudo-valuation $x \to |x|_m$ is introduced on the field of rational numbers $\mathbb{Q}$. If n is a natural number, then it can be presented in the form $n = m^r k$, where k is prime with respect m Now $|n|_m = m^{-r}$ (by standard way it is extended to $\mathbb{Q}$). The only difference from the p-adic case is that the condition (1.2) is not valid for $|\cdot|_m$, when m is not a prime number. For example, let $m = 4$, then $|2|_4 = 1$ but $|2 \times 2|_4 = 1/4 \neq |2|_4|2|_4$. Instead of (1.2), the weaker condition

$$|xy|_m \leq |x|_m|y|_m$$

holds in a general case. Such function is called *a pseudo-valuation.* A completion of the field of rational numbers $\mathbb{Q}$ with respect to the metric corresponding to the pseudo-valuation $|\cdot|_m$ is a set of m-adic numbers $\mathbb{Q}_m$. A generalization of the canonical expansion (3.5) holds also here (to exchange p and m in (3.5) ; digits belong to the set $\{0, 1, ..., m-1\}$). This canonical expansion is also unique to differ from the real case. Algebraic operations can be realized on $\mathbb{Q}_m$ with the aid of canonical expansion. No problem to check up that $\mathbb{Q}_m$ is a ring. But there are problems with division.

Example 3.1 Let $m = 6$, and let us determine two 6-adic numbers $a = 2a_1a_2a_3.....$ and $b = 3b_1b_2b_3...$ such that $ab = 0$. There is a possibility to resolve the last infinite system of equations on digits of a and b step by step. For example, after three steps we get $a = 2101...$ and $b = 3120...$

Thus, zero divisors exist in $\mathbb{Q}_m$, where m is not a prime number. That is why *it is not a field, but only a ring.* In the same way the ring of m-adic integers $\mathbb{Z}_m$ is introduced. It is compact and the set of natural numbers is dense in $\mathbb{Z}_m$. The ring $\mathbb{Q}_m$ is complete and separable by the definition. Its metric is an ultrametric. $\mathbb{Q}_m$ is zero-dimensional, totally disconnected topological space.

Let us pay attention to the case when $m = p^r, r = 1, 2, ...$ be a power of a prime. You can see that every canonical m-adic expansion can be considered as a canonical expansion of a p-adic number. From the other hand, every p-adic canonical expansion can be rewritten as a m-adic one. So $\mathbb{Q}_{p^r}$ and $\mathbb{Q}_p$ coincide as sets. There is no problem to see that their topologies are equivalent. Thus, $\mathbb{Q}_{p^r} = \mathbb{Q}_p$. These result can be generalized. Let $p_1, p_2, ..., p_k$ be distinct primes, and let $m = p_1^{r_1}...p_k^{r_k}$ and $l = p_1^{s_1}...p_k^{s_k}$, where $r_1, ..., s_k$ are the positive integers. Repeating previous considerations, we get $\mathbb{Q}_m = \mathbb{Q}_l$ as topological rings. Thus *all m-adic completions can be obtained by just considering the cases where* $m = p_1...p_k$ *is a product of one or more distinct primes.* The following basic result was proposed by K. Hensel.

Theorem 3.1. *Let* $p_1, ..., p_k$ *be* k *distinct primes and* $m = p_1...p_k$. *Then the ring* $\mathbb{Q}_m$ *is a direct sum of p-adic fields* $\mathbb{Q}_{p_i}$:

$$\mathbb{Q}_m = \mathbb{Q}_{p_1} \oplus ... \oplus \mathbb{Q}_{p_k}.$$

We note that p-adic fields $\mathbb{Q}_p$ and $\mathbb{Q}_q$ are not isomorphic if $p \neq q$. The field of real numbers $\mathbb{R}$ is not isomorphic to any $\mathbb{Q}_p$.

For us it is also important that it is not possible to define a partial order structure $\geq$ on $\mathbb{Q}_p$ satisfying (i) $1 \geq 0 \geq -1$; (ii) if $a \geq 0, b \geq 0$ then $a + b \geq 0$; (iii) if $a_n \geq 0$ for all n, $lim_{n \to \infty} a_n = a$ then $a \geq 0$.

For example,

$$\sum_{n=1}^{\infty} nn! = \sum_{n=1}^{\infty} (n+1)n! - \sum_{n=1}^{\infty} n! = \sum_{n=1}^{\infty} (n+1)! - \sum_{n=1}^{\infty} n! = -1,$$

in every $\mathbb{Q}_p$.

4. Extensions of Non-Archimedean Fields

1. On a structure of quadratic extensions and algebraic closure of non-Archimedean fields. We are interested in a p-adic analogue of complex numbers.

As we know the field of complex numbers $\mathbb{C}$ is the quadratic extension of $\mathbb{R}$ with the aid of the equation $x^2 + 1 = 0 : \mathbb{C} = \mathbb{R}(i), i = \sqrt{-1}, z = x + iy, x, y \in \mathbb{R}$. In this case we have a very simple algebraic structure because this quadratic extension is in the same time the algebraic closure of the field of real numbers (every polynomial equation has a solution in $\mathbb{C}$).

In the p-adic case there is no such simple structure. A quadratic extension is not unique as in the real case. If $p = 2$, then there are seven different quadratic extensions and if $p \neq 2$ then there are three different quadratic extensions. Thus, if we consider a fixed quadratic extension $\mathbb{Q}_p(\sqrt{\tau})$ of $\mathbb{Q}_p$, then there exist such p-adic numbers for which it is impossible to find a square root in $\mathbb{Q}_p(\sqrt{\tau})$. All these quadratic extensions are not algebraically closed. And extensions of any finite order are not algebraically closed. The algebraic closure of $\mathbb{Q}_p$ is constructed as an infinite chain of the extensions of finite orders. That is why, it is an infinite dimensional vector space over $\mathbb{Q}_p$. And this algebraic closure is not a complete field. So we must consider the completion of this field. And it is the final step of this long procedure because this completion is algebraically closed field. Let us denote this field by $\mathbb{C}_p$. In mathematical literature this field is called the field of complex p-adic numbers. But there is a problem to use $\mathbb{C}_p$ as a field of complex p-adic numbers in applications to physics.

As we know in usual quantum mechanics a great role plays the automorphism $inv : \mathbb{C} \to \mathbb{C}, z \to \bar{z}$. The real numbers field is invariant under the action of this automorphism as the expression $|z|^2 = z\bar{z}$ which is considered in the quantum mechanics as a probability. This automorphism is connected with the charge of the quantum particle. The action of inv changes the sign of the charge of the particle and in the quantum field theory this automorphism corresponds to the process *particle* $\to$ *antiparticle.* As the probability is an invariant of inv there is a particle-antiparticle symmetry in quantum physics. That is why we need an analogue of inv in p-adic valued quantum mechanics.

But there is no such an involution automorphism $inv_p : \mathbb{C}_p \to \mathbb{C}_p$. The simplest possibility in this direction is to use a quadratic extension of $\mathbb{Q}_p$ with an analogue of complex conjugation automorphism. Mainly we shall use quadratic extensions of $\mathbb{Q}_p$ as the base of our physical construction. But we shall also use $\mathbb{C}_p$ to describe some quantum models where will be a possibility to work without an involution.

Let us suppose that the quadratic equation $x^2 - \tau = 0$, $\tau \in K$, has no solution in the field K. We use the symbol Z for the quadratic extension $K(\sqrt{\tau})$ of the field K. The elements of Z are represented as $z = x + \sqrt{\tau}y$, x, $y \in K$, the conjugation operation $\bar{z} = x - \sqrt{\tau}y$, the valuation on Z is also denoted by $|\cdot|_K$, $|z|_K = \sqrt{||z|^2|_K}$. Introduce also an analogue of the Euclidean square-length $|z|^2 = z\bar{z} = x^2 - \tau y^2$. This square-length assumes its values in the field K (but $|\cdot|_K$ assumes its values in the field of real numbers).

2. Squares in the field of p-adic numbers. Our main interest in a non-Archimedean valued quantum mechanics and field theory is the p-adic case. This is why we must consider this case more carefully.

Using the canonical expansion of a p-adic number $x \neq 0$, we can write it as $x = p^m\epsilon$, where $|\epsilon|_p = 1$ and m is an integer. If x is a square of a p-adic number $y = p^k\epsilon_0$, $|\epsilon_0|_p = 1$, then $m = 2k$ and $\epsilon = \epsilon_0^2$. Thus, to describe all squares in $\mathbb{Q}_p$, it is sufficient to describe all elements ϵ, $|\epsilon|_p = 1$, which are squares, $\epsilon = \epsilon_0^2$.

Theorem 4.1. *Let $p \neq 2$. A p-adic number*

$$\epsilon = c_0 + c_1 p + c_2 p^2 + ..., c_i = 0, ..., p-1, c_0 \neq 0,$$

is a square iff c_0 is a square residue mod p.

Proof. If $\epsilon = \eta^2$ and $\eta = b(\bmod\ p)$, where b is an integer, then $c_0 = b^2(\bmod\ p)$. On the other hand, let $c_0 = b^2(\bmod\ p)$. We consider the polynomial $F(x) = x^2 - \epsilon$. We have $F(b) = 0(\bmod\ p)$ and $F'(b) = 2b \neq 0(\bmod\ p)$. Using an analogue of the Newton method, see Appendix , we get that there exists such $\eta \in Z_p$ that $F(\eta) = 0$ and $\eta = b(\bmod\ p)$. Thus, $\epsilon = \eta^2$ and the theorem is proved.

Corollary 4.1. *If $p \neq 2$, then every element ϵ such that $\epsilon =$* $1(\bmod\ p)$ *is a square in $\mathbb{Q}_p$.*

Let us consider a multiplicative subgroup $\mathbb{Q}_p^*$ of $\mathbb{Q}_p$ and a set $\mathbb{Q}_p^{*2} = \{u = v^2 : v \in \mathbb{Q}_p^*\}$ of all squares in $\mathbb{Q}_p^*$.

Corollary 4.2. *If $p \neq 2$, then the index $(\mathbb{Q}_p^* : \mathbb{Q}_p^{*2})$ is equal to 4.*

Proof. Indeed, if $|\epsilon|_p = 1$ and ϵ is not a square, then a ratio of every two numbers in $(1, \epsilon, p, \epsilon p)$ is not a square in $\mathbb{Q}_p$. On the other hand, it is not a problem to prove that every nonzero p-adic number may be presented as a product of one of the numbers $(1, \epsilon, p, \epsilon \cdot p)$ and a square.

For example, let us consider the case $p = 3$. Then a 3-adic number $x \neq 0$ is a square iff

$$x = 3^{2k}(1 + c_1 3 + c_2 3^2 + ...).$$

Thus, in particular, a square root $\sqrt{2}$ does not exist in Q_3. We can choose $\epsilon = 2$ and then (1, 2, 3, 6) are four numbers of the previous corollary. There is no problem to see directly that every 3-adic number x may be presented in the form of a product of one of these numbers and a square. It suffices to study the case $|x|_3 = 1$. If x is not a square, then $x = 2 + c_1 3 + c_2 3^2 + ... = 2(1 + c'_1 3 + c'_2 3^2 + ...) = 2w$, where w is a square.

There are 3 different quadratic extensions of $\mathbb{Q}_p$ if $p \neq 2$. These are $\mathbb{Q}_p(\sqrt{\epsilon}), \mathbb{Q}_p(\sqrt{p}), \mathbb{Q}_p(\sqrt{\epsilon \cdot p})$. In particular, if $p = 3$, we can choose the realizations $Q_3(\sqrt{2}), Q_3(\sqrt{3}), Q_3(\sqrt{6})$. If p=5, then there do not exist square roots of $\epsilon_1 = 2$ and $\epsilon_2 = 3$. Thus, we can choose $Q_5(\sqrt{2})$, $Q_5(\sqrt{5})$, $Q_5(\sqrt{10})$ as all quadratic extensions of Q_5. Another possibility is to choose $Q_5(\sqrt{3}), Q_5(\sqrt{5}), Q_5(\sqrt{15})$. But $3 = 2(3/2) = 2(4 + 2 \cdot 5 + ... + 2 \cdot 5^n + ...) = 2v$, where $v = w^2$, and $\sqrt{3} = \sqrt{2}w$. Hence, $Q_5(\sqrt{2}) \cong Q_5(\sqrt{3})$ and $Q_5(\sqrt{10}) \cong Q_5(\sqrt{15})$.

Now let us study the case $p = 2$.

Theorem 4.2. *Let $p = 2$. A 2-adic number ϵ, $|\epsilon|_2 = 1$, is a square iff $\epsilon = 1(\mathrm{mod}\ 8)$.*

Proof. Necessity is a consequence of the fact that a square of an odd number is always compared with $1(\mathrm{mod}\ 8)$. On the other hand, we consider a polynomial $F(x) = x^2 - \epsilon$. Then $F(1) = 0(\mathrm{mod}\ 8)$ and $F'(1) = 2 \neq 0(\mathrm{mod}\ 4)$. We again apply Theorem 2.1 of Appendix (choosing $\delta = 1$).

Corollary 4.2. *Index $(Q_2^* : Q_2^{*2})$ is equal 8.*

Proof. The unit sphere $S_1 \equiv S_1(0)$ is a multiplicative subgroup in Q_2^*. Let us denote by S_1^2 the subgroup of squares in S_1. On the basis of Theorem 4.2, we get that the numbers $A = (1, 3, 5, 7)$ are the representatives of the residue classes of S_1/S_1^2. To show this fact, we need only to prove that every element $x \in S_1$ may be presented in the form $x = au^2, a \in A, u \in S_1$. As $x = a + x_3 \cdot 2^3 + ...$, where $a \in A$, then $x = a(x/a)$. But $1/3 = 11(01), 1/5 = 1011(011), 1/7 = 111(011)$.

We use the canonical expansion, where brackets are used to show a period. Hence if $a = 3$, then $x/3 = 100x_3...$ is a square; if $a = 5$, then $x/5 = 100x_3...$ is a square and if $a = 7$, then $x/7 = 100x_3...$ is a square. Hence there exists $\sqrt{x/a}$. If we join to A the products $2 \cdot 1, 2 \cdot 3, 2 \cdot 5, 2 \cdot 7$, we get the complete system of representatives of the residue classes of Q_2^* with respect to Q_2^{*2}.

We shall interested (in applications to the problem of the quantization) in square roots of $x = -1$ in $\mathbb{Q}_p$. Denote one of these square roots by the symbol i_p (another is equal to $-i_p$. There is no problem to show on the basis of Theorems 4.1 and 4.2 that this square root exists in $\mathbb{Q}_p$ if $p = 1(\bmod 4)$, see also [17]. For example, if $p = 5$ then $i_5 = 21213423032204132404...$ (there we use the convention of Section 3 on the canonical expansion of p-adic integers).

3. The field of complex p-adic numbers $\mathbb{C}_p$. We shall use not only quadratic extensions but also so called field of complex p-adic numbers $\mathbb{C}_p$. To build this field, we consider an algebraic closure $\mathbb{Q}_p^a$ of the field $\mathbb{Q}_p$ (the minimal field extension of $\mathbb{Q}_p$, where every polynomial $p(x)$ with p-adic coefficients has a root). There is the unique possibility to extend the valuation $|\cdot|_p$ from $\mathbb{Q}_p$ to $\mathbb{Q}_p^a$. We shall never use the concrete form of this continuation of $|\cdot|_p$. This is why we do not pay attention to this procedure of continuation. The same symbol $|\cdot|_p$ will be used to denote the valuation on $\mathbb{Q}_p^a$. But a new difficult arises on this step. The metric space $\mathbb{Q}_p^a$ is not complete. The completion $\mathbb{C}_p$ of $\mathbb{Q}_p^a$ is said to be a *field of complex p-adic numbers.*

4. Representation of p-adic numbers in the form $a = x^2 - \tau y^2$. Signum-functions on the field of p-adic numbers. Now we discuss the following problem. Let $Z = \mathbb{Q}_p(\sqrt{\tau})$ be the fixed quadratic extension of $\mathbb{Q}_p$. We are interested in elements $a \in \mathbb{Q}_p$, which can be presented as $a = |z|^2 = z\bar{z} = x^2 - \tau y^2$. It is useful to study more general problem of representations

$$a = x^2 - \alpha y^2, \tag{4.1}$$

where α is an arbitrary p-adic number, $\alpha \neq 0$. Denote a set of all such nonzero numbers by H_α. This set is a multiplicative group. Indeed, if

$b = x^2 - \alpha y^2$ and $b_1 = x_1^2 - \alpha y_1^2$, then

$$bb_1 = (xx_1 + \alpha yy_1)^2 - \alpha(xy_1 + yx_1)^2, b^{-1} = (\frac{x}{b})^2 - \alpha(\frac{y}{b})^2.$$

Theorem 4.3. *If α is a square in $\mathbb{Q}_p$, then $H_\alpha = \mathbb{Q}_p^*$. If α is not a square in $\mathbb{Q}_p$, then the index $(\mathbb{Q}_p^* : H_\alpha) = 2$.*

The proof can be found in [17].

Now let τ be the fixed element, which is not a square. Let us introduce a function $\text{sign}_\tau(x) = 1$ if $x \in H_\tau$ and $\text{sign}_\tau(x) = 0$ if $x \notin H_\tau$. It is a natural generalization of the usual sign-function on $\mathbb{R}$ (if we realize it with the aid of its quadratic extension C)). As H_τ is a multiplicative subgroup of $\mathbb{Q}_p^*$ with the index 2, then sign_τ is a multiplicative function on $\mathbb{Q}_p^*$, $\text{sign}_\tau(ab) = \text{sign}_\tau(a)\text{sign}_\tau(b)$.

5. Normed and Locally-Convex Spaces

Let E be a linear space over K. The *non-Archimedean norm* on E is a mapping $|\cdot| : E \to \mathbb{R}_+$, satisfying the following conditions: (1)$||x|| = 0 \iff x = 0$, (2) $||\lambda x|| = |\lambda|_K ||x||, \lambda \in K$, (3) $||x + y|| \leq \max(||x||, ||y||)$. If $||\cdot||$ satisfies conditions 2 and 3, then $||\cdot||$ is an *non-Archimedean prenorm.* As usual, we define a non-Archimedean Banach space B as a complete normed space over K. The metric $\rho(x, y) = ||x - y||$ is an ultrametric. Hence, every non-Archimedean Banach space is a zero-dimensional and totally disconnected. All balls $U_r(a) = \{x \in B : ||x - a|| \leq r\}$ and $U_r^-(a) = \{x \in B : ||x - a|| < r\}, a \in B, r \in \mathbb{R}_+$, are clopen. A dual space B' is defined as the space of continuous K-linear functionals $l : B \to K$. Let us introduce the usual norm on B' :

$$||l|| = \sup_{x \neq 0} \frac{|l(x)|_K}{||x||}.$$

The space B' endowed with this norm is a Banach Space.

We shall use in the theory of distributions the simple generalization of Theorem 1.2:

Theorem 5.1 *Let B be a non-Archimedean Banach space. The series $\sum_{n=1}^{\infty} a_n, a_n \in B$, converges in B if and only if $a_n \to 0, n \to \infty$.*

The proof is also a consequence of the strong triangle inequality.

If the topology on E is defined by a system of non-Archimedean prenorms, then E is a non-Archimedean locally convex space. Neighbourhoods of a point a have the form $U_{\alpha,r}(a) = \{x \in E : ||x - a||_\alpha \leq r\}, r \in \mathbb{R}_+, ||\cdot||_\alpha$ is one of the prenorms defining topology on E. The set of these prenorms is denoted by the symbol P_E. In particular, a sequence $\{x_n\}$ of elements of E converges to $x \in E$ if $||x - x_n||_\alpha \to 0, n \to \infty$, for every $||\cdot||_\alpha \in P_E$. In standard way we define a Fréchet space E as a locally convex space which topology is equivalent to metric topology and E is complete with respect to this metric topology. A complete locally convex space E is a Fréchet space if and only if its topology is defined by a countable set of prenorms $\{||\cdot||_n\}_{n=1}^{\infty}$.

Non-Archimedean Banach and locally convex algebras are also defined as usual (the operation of the multiplication is continuous). If B is a Banach algebra the following inequality holds: $||a\, b|| \leq C||a||\, ||b||$; if E is a locally convex algebra then for every prenorm $||\cdot||_\alpha \in P_E$ there exists such prenorm $||\cdot||_\beta \in P_E$ that the following inequality holds: $||a\, b||_\alpha \leq C||a||_\beta\, ||b||_\beta$.

Remark 5.1. As we know, in a real case a locally convex space is defined as a topological vector space in which there exists a fundamental system of neighbourhoods of zero consisting of convex sets. The existence of such a system of neighbourhoods is equivalent to the existence of a fundamental system consisting of absolutely convex (convex balanced) subsets. In a non-Archimedean case the concept of an absolute convexity is more natural than that of a convexity, i.e., $A \subset E$ is absolutely convex if $\lambda x + \mu y \in A$ for $x, y \in A$, $\lambda, \mu \in K$, $|\lambda|_K$, $|\mu|_K \leq 1$. There is an ordinary correspondence between the absolutely convex subsets of the non-Archimedean linear space E and the non-Archimedean prenorms, E is locally convex if and only if it includes a fundamental system of neighbourhoods consisting of absolutely convex subsets. For the non-Archimedean theory of absolutely convex spaces see [86, 90 − −92]. But this language will not be used in the book. We shall use only prenorms description of topologies.

Remark 5.2. In the ordinary theory absolutely convex compact subsets in locally convex spaces play an important part. In a non-Archimedean case, a direct generalization leads to a number of pathological examples and that is why a new class of objects, compactoids, is introduced here [91, 92], which are used to develop the theory [91, 92].

The simplest example of a non-Archimedean Banach space is the space $K^n = K \times ... \times K(n$ times) with the non-Archimedean norm $||x|| = \max_{1<j<n} |x_j|_K$.

6. Continuous, Differentiable and Analytic Functions

1. Locally constant functions. Let $O \subset K$. The symbol $C(O) \equiv C(O, K)$ denotes the K-linear space of continuous functions $f : O \to K$. The symbol $C_b(O) \equiv C_b(O, K)$ denotes the subspace consisting of bounded functions. The uniform norm $||f|| = \sup_{x \in O} |f(x)|_K$ is introduced on $C_b(O)$. It is a non-Archimedean Banach space. If O is compact then $C(O) = C_b(O)$ and for $f \in C(O)$ we have $||f|| = \max_{x \in O} |f(x)|_K$.

A basic property of continuous functions in the non-Archimedean case is the fact that they can uniformly be approximated by locally constant functions.

Let $O \subset K$. By the definition a function $f : O \to K$ is locally constant if for each $x \in O$ there is a neighbourhood U of x such that f is constant on $U \cap O$.

A subset $U \in O$ is called relatively open, close, clopen, ... if it is open, close, clopen, ... with respect to the topology of O induced from K. The (K-valued) characteristic function $\phi_{U,O}(x)$ of a relatively clopen subset $U \subset O$ (defined as $\phi_{U,O}(x) = 1$ if $x \in U$ and $\phi_{U,O}(x) = 0$ if $x \in O \setminus U$) is locally constant. If $f : O \to K$ is locally constant then O admits a partition into relatively clopen sets U_i, where i runs through some indexing set, such that f is constant on U_i for each i. Locally constant functions are continuous. The locally constant functions on O form a K-linear subspace of $C(O)$.

Proposition 6.1. *Let $O \subset K$ and $f \in C(O), \epsilon > 0$. Then there is a locally constant function $g : O \to K$ for which $|f(x) - g(x)|_K < \epsilon$ for all $x \in O$. The bounded locally constant functions form a dense subspace of $C_b(O)$.*

Proof. We only need to prove the first statement. The relation $\sim$ on O defined by $x \sim y$ if $|f(x) - f(y)|_K < \epsilon$ is an equivalence relation whose equivalence classes $U_i (i \in I)$ are relatively clopen. For each $i \in I$ choose an element $a_i \in U_i$ and define $g : O \to K$ by equality $g(x) = f(a_i)$ if $x \in U_i, i \in I$. Then g is constant on each U_i and $|f(x) - g(x)|_K < \epsilon$ for all $x \in O$.

2. Approximation by polynomials. We shall also use an approximation of continuous functions by means of polynomial functions $q(x) = \sum_{k=1}^{N} a_k x^k$, where $a_k \in K$. The following theorem will play an important role in non-Archimedean valued integration and probability theories. But the proof will not be used everywhere. So it is possible to see only the formulation of Theorem.

Theorem 6.1. (Kaplansky) *Let T be a compact subset of K, let $f \in C(T)$ and $\epsilon > 0$. Then there is a polynomial function $q : K \to K$ such that $|q(x) - f(x)|_K < \epsilon$ on T.*

Proof. By previous theorem it suffices to solve the problem for locally constant function f. By compactness there is a $\delta \in (0, 1)$ such that f is constant on each of (finitely many) balls in T of radius δ. So we may even assume that f is the characteristic function of a ball (in T) of radius δ. Without loss of generality, let $0 \in T, f(0) = 1$. Choose $c_1, ..., c_m \in T$ such that $T \subset U_\delta(0) \cup U_\delta(c_1) \cup ... \cup U_\delta(c_m)$, where these balls (in K) are pairwise disjoint and $|c_1|_K \leq |c_2|_K ... \leq |c_m|_K$. Then $\delta < |c_1|_K$; choose natural s such that $(\delta/|c_1|_k)^s < \epsilon$. Inductively we shall define integers $n_1, ..., n_m$ such that the polynomial function defined by

$$q(x) = \prod_{j=1}^{m} (1 - (c_j^{-1} x)^s)^{n_j},$$

$x \in K$, is well for Theorem. It is the same to show that $|q(x) - 1|_K < \epsilon$ for $x \in U_\delta(0)$ and $|q(x)|_K < \epsilon$ for $x \in U_\delta(c_1) \cup ... \cup U_\delta(c_m)$. First let $x \in U_\delta(0)$. Then $[1 - (c_j^{-1} x)^s] \in U_\epsilon(1)$ for all j. Since $U_\epsilon(1)$ is a multiplicative group (at least if $\epsilon < 1$) we may conclude that $q(x) \in U_\epsilon(1)$, i.e.$|q(x) - 1|_K < \epsilon$. This result does not depend on the choice of $n_1, .., n_m$. Now let $x \in U_\delta(c_i)$ for some $i = 1, 2, ..., m$. Then $|x - c_i|_K \leq \delta$ and $|x|_K = |c_i|_K$ so that

$$|1 - c_j^{-s} x^s|_K \leq \max(1, |x/c_j|_K^s) \leq |c_i/c_j|_K^s \quad (j < i)$$

$$|1 - c_i^{-s} x^s|_K \leq |1 - c_i^{-1} x|_K \leq \delta |c_i^{-1}|_K \leq \delta |c_1^{-1}|_K$$

$$|1 - c_j^{-s} x^s|_K \leq \max(1, |x/c_j|_K^s) \leq 1 \quad (j > i)$$

In order that $|q(x)|_K \leq \epsilon$ we need

$$|c_i/c_1|_K^{sn_1} |c_i/c_2|_K^{sn_2} ... |c_i/c_{i-1}|_K^{sn_{i-1}} (\delta/|c_1|_K)^{n_i} < \epsilon$$

If $n_1, .., n_{i-1}$ are already chosen we can, since $\delta/|c_1|_K < 1$, always choose n_i such that this inequality holds.

Kaplansky's theorem holds for arbitrary K whereas Weierstrass' approximation theorem becomes a falsity if we replace $\mathbb{R}$ by $\mathbb{C}$. If K is locally compact, in particular $\mathbb{Q}_p$, Kaplansky's theorem can be applied to balls $T = U_r(a)$.

3. Differentiable functions with zero derivative. A definition of a derivative is presented in the standard way. Let W be an open subset of K and $a \in W$. A function $f : W \to K$ is differentiable at a if the derivative $f'(a)$ of f at a

$$f'(a) = \lim_{h \to 0} [f(x + h) - f(x)]/h$$

exists. f is differentiable on W if it is differentiable in every point $a \in W$.

The main difference from the Archimedean case ($\mathbb{R}$ or $\mathbb{C}$) is the existence of functions which are not constants or locally constants and has the zero derivative.

Example 6.1. There is an injective $f : \mathbb{Z}_p \to \mathbb{Z}_p$ whose derivative is 0. For $x = a_0 + ... + a_n p^n + ... \in \mathbb{Z}_p$ set

$$f(x) = a_0 + a_1 p^2 + ... + a_n p^{2n} + ...$$

We show that this function satisfies the requirements. Let $x = a_0 a_1 ... a_n ...$ and $y = b_0 b_1 ... b_n ...$ be elements of $\mathbb{Z}_p$ and $|x - y|_p = p^{-j}$ for some $j = 0, 1, ...$ Then $a_0 = b_0, ..., a_{j-1} = b_{j-1}$ and $a_j \neq b_j$. It follows that $|f(x) - f(y)|_p = p^{-2j}$. So

$$|f(x) - f(y)|_p = |x - y|_p^2, \ (x, y \in \mathbb{Z}_p)$$

and this is why $f' = 0$.

But we shall not be interested in such unusual and complicated moments in non-Archimedean analysis. Analytic (and continuous) functions are the main objects of our considerations.

4. Power series and analytic functions. Let us consider a power series $\sum_{n=1}^{\infty} c_n x^n$, where x and coefficients $c_j \in K$. The region of convergence of a series is the set of all x for which this series converges.

The radius of convergence is defined by

$$\rho = \left(\overline{\lim}_{n\to\infty} |c_n|_K^{1/n}\right)^{-1}.$$

The radius of convergence has the usual properties as for $\mathbb{C}$. The series converges on the set $\{x \in K : |x|_K < \rho\}$ and diverges on the set $\{x \in K : |x|_K > \rho\}$. For each $\tau, 0 < \tau < \rho$, convergence is uniform on $\{x \in K : |x|_K \leq \tau\}$. The function

$$x \to \sum_{n=0}^{\infty} c_n x^n \quad (|x|_K < \rho)$$

is differentiable. Its derivative is computed in the standard way. Contrary to the complex case it is not always true that for a power series there is a unique $r \in [0, \infty]$ such that series converges if $|x|_K < r$ and diverges if $|x|_K > r$. So ρ is not defined with the aid of this property. Such situation can be realized for $K = \mathbb{Q}_p$ as the valuation assume discrete values p^l. For example, let $\rho = p$ and the region of convergence is the ball $U_p(0)$, so the series converges also on the sphere $S_p(0)$. Now we can choose each $r \in [p, p^2]$.

The problem of a convergence of a power series on the boundary plays a large role in complex analysis. But an analogous problem has a simple structure in the non-Archimedean case. Let $\rho < \infty$, $\rho \neq 0$. Then a series converges *either everywhere or nowhere on the sphere* $S_\rho(0)$. But this sphere is not the boundary of the ball $\{x : |x|_K < \rho\}$. It is also possible, for example, in $\mathbf{C}_p$, that there exists such series that the set $S_\rho(0)$ is empty.

Hence the region of convergence is a ball $U_\rho(0)$ or a ball $U_\rho^-(0)$.

There is also difference between Archimedean and non-Archimedean cases in the convergence of a series and its formal derivative series. The regions of convergence may differ.

The same considerations can be presented for power series $\sum c_n(x-a)^n$ where $a \in K$. There the region of convergence is a ball $U_\rho(a)$ or a ball $U_\rho^-(a)$. Let O be one of these balls. A function $f : O \to K$ is analytic if it can be expanded into a power series

$$f(x) = \sum_{n=0}^{\infty} c_n(x-a)^n, \quad c_n \in K \tag{6.1}$$

This definition is the same as in the complex case. But we get a surprising difference from complex case by one step. As we know every point $b \in O$ can be considered as the centre of these ball. Does the definition of analytic function on O depend on the point of the expansion ? We understand intuitively that it can not depend. And it is really so. If f can be expanded into a power series on O with respect to one point a, then it can also be expanded on O with respect to any point $b \in O$. The procedure of the reexpansion of the series (6.1) in a new point b is the same to the complex case. To compute new coefficients the binomial expansion of $(x-a)^n = [(x-b)+(b-a)]^n$ is used as usual. But we need to use the strong triangle inequality to prove the convergence of a new series on O. This result can be considered as *no analytic continuation result.* If we expand the function with respect to a new point, we get again the same region of convergence. Thus in the non-Archimedean case all the theory of analytic functions is reduced to the theory on balls. The centre of the ball does not play a great role. This is why we shall restrict our considerations to analytic functions on balls with zero centre.

As a ball $U_r(a)$ in K^n is a direct product of one-dimensional balls of the same radius, there is no problem to generalize all considerations to a case of many variables. Let O be a ball in K^n. A function $f : O \to K$, where O is a ball in K^n, is analytic if it can be expanded into a power series on O. We shall continue our analytic consideration at the beginning of the next chapter to introduce main spaces of analytic functions on balls in K^n (with zero centre).

5.Exponent. Exponents play an important part in the construction of the non-Archimedean analysis. The exponent in K is defined by the series

$$e^x = \sum_{n=0}^{\infty} x^n/n!$$

Using Theorem 1.2. and the exponential estimate (1.5) of the function $1/|n!|_K$, we find that the series for the exponents converges if $|x|_K < b^{-1}$. In particular, in p-adic case it converges if $|x|_p < p^{1/(1-p)}$.

7. The Mahler Theory of an Integration on the Ring of p-adic Integers

1. The Mahler expansion of continuous function. We study continuous functions $f : \mathbb{Z}_p \to \mathbb{Q}_p$. The set $\mathbb{Z}_p$ is a compact and this is why each continuous function f is the uniformly continuous. So for each natural s there exists a natural $t = t(s)$ such that, if $|x - y|_p \leq p^{-t}$, then $|f(x) - f(y)|_p| \leq p^{-s}$. As it was discussed the set of natural numbers is a dense subset in $\mathbb{Z}_p$. Hence, each continuous function f is determined by means to its values in natural points: $f(0), f(1), ..., f(n), ...$

Now let us form the expressions

$$a_n = \sum_{k=0}^{n} (-1)^k C_n^k f(n-k) \ (n = 0, 1, 2, ...) \tag{7.1}$$

where $C_n^k = n!/[k!(n-k)!] = n(n-1)(n-2)...(n-k+1)/k!$ are the binomial coefficients. The expressions (7.1) are standard in the difference calculus. The formal interpolation series is defined by

$$f^*(x) = \sum_{n=0}^{\infty} a_n C_x^n, \tag{7.2}$$

where $C_x^n = x(x-1)(x-2)...(x-n+1)/n!$.

First of all let us consider this interpolation series on the set of natural numbers: $x = N$. Then $C_N^n = 0$ for $n > N$ and $f^*(N)$ is the finite sum

$$f^*(N) = \sum_{n=0}^{N} a_n C_N^n. \tag{7.3}$$

Using the properties of binomial coefficients, it is possible to prove that $f^*(N) = f(N)$. These are the standard computations of difference calculus which are not connected in any way to p-adios (input expressions for the coefficients a_n into (7.1), multiply binomial coefficients and use the binomial expansion to compute the sum of binomial coefficients).

Thus, *$f(x)$ and $f^*(x)$ coincide on the set of natural numbers.* This is why, if we show that $f^*(x)$ is convergent for all x and the sum is a continuous function then $f^*(x) \equiv f(x)$. But the proof is very long and we prefer only to present this result as the theorem.

Theorem 7.1. (Mahler) *Let $f : \mathbb{Z}_p \to \mathbb{Q}_p$ be a continuous function and interpolation coefficients a_n are computed with the aid of (7.1). Then*

1)$\lim_{n\to\infty} a_n = 0$;

2)the interpolation series (7.2) converges uniformly on $\mathbb{Z}_p$ to $f(x)$;

3)$||f|| = \max\{|a|_n : n \in \{0, 1, 2, ...\}\}$.

The inverse statement is also valid: *if $\{a_n\}$ is a null sequence in $\mathbb{Q}_p$ then the series (7.2) defines a continuous function on $\mathbb{Z}_p$.*

2. A measure as a functional. We define a $\mathbb{Q}_p$-valued measure on $\mathbb{Z}_p$ as a continuous $\mathbb{Q}_p$-linear functional $\mu : C(\mathbb{Z}_p) \to \mathbb{Q}_p$. So the set of measures coincides with the dual space $C'(\mathbb{Z}_p)$. Using the Mahler expansion, we get

$$\mu(f) = \mu(\sum_{n=0}^{\infty} a_n C_x^n) = \sum_{n=0}^{\infty} a_n \mu_n, \tag{7.4}$$

where $\mu_n = \mu(C_x^n)$. Hence $|\mu(f)|_K \leq L_\mu \max_n |a_n|_k = L_\mu||f||$ where $L_\mu = \sup_n |\mu_n|_K$ (we have used Theorem 7.1). Thus, $||\mu|| \leq L_\mu$. But $||C_x^n|| = 1$ (again use Theorem 7.1 item 3) and

$$|\mu_n|_K = |\mu(C_x^n)|_K = |\mu(C_x^n)|_K/||C_x^n|| \leq ||\mu||$$

So we have proved that $||\mu|| = L_\mu$. And *every measure on $\mathbb{Z}_p$ is determined by a bounded sequence $\{\mu_n\}$ of p-adic numbers.* The inverse is also true:*every bounded sequence of p-adic numbers $\{\mu_n\}$ determines the measure μ with the aid of (7.4).*

3. A measure as a set-function. Let us denote by $\Phi(\mathbb{Z}_p)$ the system of clopen subsets of $\mathbb{Z}_p$. Characteristic function $\phi_U(x)$ is continuous for each $U \in \Phi(\mathbb{Z}_p)$. Hence it is possible to correspond a set-function $U \to \mu(U) = \mu(\phi_U)$ to each measure μ. As, in particular, the linear functional μ is additive we get that this set-function is additive

$$\mu(U \cup V) = \mu(U) + \mu(V),\ U \cap V = \emptyset. \tag{7.5}$$

As the linear functional μ is continuous we get that the set-function has the property of the boundedness

$$\sup\{|\mu(V)|_p : V \subset U, V \in \Phi(\mathbb{Z}_p)\} < \infty \tag{7.6}$$

for every $U \in \Phi(\mathbb{Z}_p)$. It can be possible to prove that every set-function $\mu : \Phi(\mathbb{Z}_p) \to \mathbb{Q}_p$ for which (7.5) and (7.6) hold can be extended to a measure on $\mathbb{Z}_p$ (continuous linear functional $\mu : C(\mathbb{Z}_p) \to \mathbb{Q}_p$). As usual the first step is to define by linearity this functional on the space E of finite linear combinations of characteristic functions of clopen subsets of $\mathbb{Z}_p$ (use (7.5)). The second step is to show that the functional $\mu : E \to \mathbb{Q}_p$ is bounded (use (7.6))

$$||\mu||_E = \sup\{|\mu(\phi)|_p/||\phi|| : \phi \in E, \phi \neq 0\} < \infty.$$

And this is why it is continuous in topology generated from $C(\mathbb{Z}_p)$. To conclude the proof we need only to apply Proposition 6.1 to extend the functional μ from E to $C(\mathbb{Z}_p)$ by continuity.So we have proved that *there is one to one correspondence between measures defined as* *linear continuous functionals on the space $C(\mathbb{Z}_p)$ and set-functions on the system $\Phi(\mathbb{Z}_p)$ of clopen subsets for which (7.5) and (7.6) hold.*

3. No Haar measure. We prove that there is no translation invariant measure $\mu : \Phi(\mathbb{Z}_p) \to \mathbb{Q}_p$. As usual a measure is called the translation invariant if

$$\mu(a+U) = \mu(U), \ a \in \mathbb{Z}_p, \ U \in \Phi(\mathbb{Z}_p). \tag{7.7}$$

At first we prove that the p^n additive cosets of $p^n\mathbb{Z}_p$ in $\mathbb{Z}_p$ form a partition of $\mathbb{Z}_p$. In view of the canonical expansion of $x \in \mathbb{Z}_p$ we get that $x = a_0 + ... + a_{n-1}p^{n-1} + z$, where $z \in p^n\mathbb{Z}_p$. Now we need only to see that the digits $a_i, 0 \leq i \leq n-1$, can be only $\{0, 1, .., p-1\}$. So there are p^n different possibilities. And different cosets have the empty intersection.

Let μ be a translation invariant measure. By (7.5) and (7.7)

$$\mu(\mathbb{Z}_p) = \sum_{i=1}^{p^n} \mu(p^n\mathbb{Z}_p) = p^n \mu(p^n\mathbb{Z}_p). \tag{7.8}$$

By (7.6) we have $\sup_n |\mu(p^n\mathbb{Z}_p)|_p < \infty$ so $\mu(\mathbb{Z}_p) = \lim_{n\to\infty} p^n\mu(p^n\mathbb{Z}_p) = 0$. In view of (7.8), we get $\mu(p^n\mathbb{Z}_p) = 0$ for every $n = 0, 1, ...$ Thus, by the translation invariance $\mu(U_r(a)) = 0$ for every ball in $\mathbb{Z}_p$ (as $r = p^{-n}, n = 0, 1, ...$). But every clopen subset is a finite disjoint union of balls.

But we note (see [101], [102]) that there exists the translation invariant measure $\mu : \Phi(\mathbb{Z}_p) \to \mathbb{Q}_q$, where $p \neq q$. This question will be discussed more careful in Ch. 6 in the context of Monna-Springer integration theory [102].

4. Volkenborn integral (uniform distribution). Using (7.8), we have that $\mu(p^n\mathbb{Z}_p) = (1/p^n)$, if $\mu(\mathbb{Z}_p) = 1$. Thus, $\mu(U_r(a)) = r$ for $r = (1/p^n)$. Hence we could define a translation invariant finite additive function on the algebra of clopen subsets, using this values of the measures of balls. But, according to the previous considerations, it would be unbounded and it would be impossible to integrate all continuous functions with respect to this "measure". There is another way to introduce this integral.

The Volkenborn integral of a function $f : \mathbb{Z}_p \to Q_p$ is

$$\int_{\mathbb{Z}_p} f(x)\mathrm{d}x = \lim_{n\to\infty}(1/p^n)(f(0) + \ldots + f(p^n - 1)).$$

Here we cannot study this integral in the rigorous way. To do this, we need to introduce a new functional space (strictly continuous differentiable functions [106]). We only note that there exist such continuous functions that the Volkenborn integral is not defined. But it is well defined for characteristic functions of balls and

$$m(U_r(a)) = \int \phi_{U_r(a)}(x)\mathrm{d}x = r, r = 1/p^n.$$

Hence the translation invariant additive set function corresponds to these integral. Then we shall use that it is well defined for every polynomial function and

$$\int_{\mathbb{Z}_p} x^n \mathrm{d}x = B_n, n = 0, 1, \ldots,$$

where B_n are the Bernoulli numbers defined with the aid of

$$a/(\mathrm{e}^a - 1) = \sum_{n=0}^{\infty} B_n a^n/n!.$$

At the moment, last equality we can consider as the equality in the sense of formal power series. The following estimate

$$|B_n|_p \le p$$

holds. The Volkenborn integral is well defined for analytical functions $f : \mathbb{Z}_p \to \mathbb{Q}_p$, $f(x) = \sum_{n=0}^{\infty} a_n x^n$, and

$$\int_{\mathbb{Z}_p} f(x)\mathrm{d}x = \sum_{n=0}^{\infty} a_n B_n.$$

The Volkenborn "measure" is in some sense a uniform distribution on the ring of the p-adic integers. Let us try to introduce a uniform distribution on the set of natural numbers N (at this moment, we consider zero as the element of N). Consider first p^n numbers $0, 1, ..., p^n - 1$, with the uniform weights $1/p^n$. Now let $n \to \infty$. We hope to get a uniform distribution. But it would be impossible to get such distribution in the usual sense. We get such distribution on $\mathbb{Z}_p$, where N is a dense subset. But it is not a measure also in the p-adic sense.

We shall study Volkenborn integral in Ch. 2 on the basis of a theory of generalized functions. A Volkenborn "measure " will be realized as a generalized function.

Chapter II

The Gauss, Lebesgue and Feynman Distributions Over Non-Archimedean Fields

The definitions of the Gauss and Feynman integrals with respect to the number line $\mathbb{R}$,

$$I_G = \int_{\mathbb{R}} \varphi(x) e^{-\frac{x^2}{2b}} \frac{dx}{\sqrt{2\pi b}}, \qquad I_F = \int_{\mathbb{R}} \varphi(x) e^{-\frac{x^2}{2bi}} \frac{dx}{\sqrt{2\pi b i}},$$

are based on the Lebesgue measure dx on $\mathbb{R}$. Gauss integral I_G is an integral with respect to the Gauss' measure on $\mathbb{R}$ which is absolutely continuous relative to the Lebesgue measure dx. The situation is more complicated for an oscillating Feynman integral. If the function $\varphi(x)$ is summable with respect to the measure dx on $\mathbb{R}$, the integral I_F is a Lebesgue integral (since the exponent is bounded). Now if φ is not Lebesgue summable, then the integral is regarded as a generalized integral in the sense of the theory of distributions on $\mathbb{R}$, i.e., $I_F = (\mu_F, \varphi)$, where μ_F is the corresponding distribution (= a generalized function) on $\mathbb{R}$.

A new problem arises in the non-Archimedean theory connected with the absence of the Lebesgue (Haar) measure dx with non-Archimedean values. The absence of the Lebesgue measure seriously complicates the construction of the non-Archimedean theory of the Gauss and Feynman integration.

The following approach to the solution of this problem was suggested in [57–59], [64], [72]. First, in the framework of the distribution theory, the Gaussian distribution is considered as a continuous functional on the space of entire analytic functions. This construction is based on Parseval's equality. *We do not know what a non-Archimedean Gaussian distribution is but, by analogy with the theory of real functions, can assume that its Laplace (Fourier) transformation is a quadratic exponent on the non-Archimedean number field.* And, by virtue of Parseval's equality, in order to see how the distribution affects the test function it is sufficient to know the Laplace (Fourier) transform of this distribution. Then the Lebesgue distribution is introduced on the non-Archimedean field as a distribution absolutely continuous relative to the Gaussian distribution.

The Parseval equality can also be used to define a non-Archimedean analog of the Feynman integral I_F. It should be pointed out that in the non-Archimedean theory there is practically no difference between the integrals I_G and I_F since the behavior of both quadratic exponents is almost the same. There is no problem here of the behavior of these exponents at the infinity since they are only defined in a certain neighborhood of zero. This is precisely the reason for the specific choice [72] of the space of test functions of the non-Archimedean argument, namely, this space consists of all functions that are analytic at zero (generally speaking, not defined throughout the field).

The idea of defining the non-Archimedean Gauss and Feynman distributions by Parseval's equality was borrowed from the theory of Feynman integrals with respect to infinite-dimensional real spaces [1], [134], [115–117], [43–45]. The Lebesgue measure dx is also absent on infinite-dimensional spaces over $\mathbb{R}$ and the problems of the ordinary infinite-dimensional analysis have much in common with non-Archimedean problems.

It is the same in the superanalysis, where there is no Lebesgue measure either but there are only distribution integrals (even in a finite-dimensional case), see [46], [49], [50]. For the Gauss and Feynman integrals with respect to non-Archimedean superspaces see [51], [66], [73], [74] (the *Volkenborn integral* over the superspace is also used here).

1. Analytic Functions over Non-Archimedean Fields

Now we introduce spaces of analytic functions which will play a basic role in a theory of distributions. For the motivations of Ch. 1 we study only the case of functions defined on balls with the centre at the zero. It is much more simple to consider only the balls of radii $R \in \Gamma$, because only such balls are the "natural balls".

Denote by $\mathcal{U}_R$ a ball $U_R(0)$ in K^n. The function $f: \mathcal{U}_R \to K$, $R \in \Gamma$, is said to be analytic if the series

$$f(x) = \sum_\alpha f_\alpha x^\alpha, \qquad f_\alpha \in K, \tag{1.1}$$

converges uniformly on $\mathcal{U}_R$. Here as usual $\alpha = (\alpha_1, ..., \alpha_n), \alpha_i = 0, 1, 2, ...,$ and $x^\alpha = x_1^{\alpha_1} \ldots x_n^{\alpha_n}$. By virtue of the non-Archimedean Cauchy criterion of convergence (1.1), it is necessary and sufficient that

$$\lim_{|\alpha| \to \infty} |f_\alpha|_K R^{|\alpha|} = 0$$

To prove that this condition is necessary we use $R \in \Gamma$. So there exists such $a_R \in K$ that $|a_R|_K = R$. The point $x_R = (a_R, ..., a_R) \in \mathcal{U}_R$. And if (1.1) converges in this point, then (1.2) holds.

The topology in the space $\mathcal{A}(\mathcal{U}_R) \equiv \mathcal{A}(\mathcal{U}_R, K)$ $(\equiv \mathcal{A}_R)$ of the functions analytic on the ball $\mathcal{U}_R$ is defined by the non-Archimedean norm $||f||_R = \max_\alpha |f_\alpha|_K R^\alpha$. This is a non-Archimedean Banach space.

The function $f: K^n \to K$ is *entire* if series (1.1) converges on the ball $\mathcal{U}_R$ for any $R \in \Gamma$. The topology in the space of entire functions $\mathcal{A}(K^n) \equiv \mathcal{A}(K^n, K)(\equiv \mathcal{A})$ is defined by a system on non-Archimedean norms $\{|| \cdot ||_R\}_{R \in R_+}$.

A sequence of entire functions $\{f_n\}$ converges in $\mathcal{A}$ if it converges uniformly on each ball $\mathcal{U}_R$.

There is no problem to see that this topology could be defined with the aid of arbitrary sequence of norms $\{||\cdot||_{R_k}\}_{k=1}^{\infty}$, where $\lim_{k\to\infty} R_k = \infty$. It is a non-Archimedean Fréchet space. Such type of topology is known as the topology of projective limit of Banach spaces

$$\mathcal{A}(K^n) = \lim_{R\to\infty} \mathrm{proj}\, \mathcal{A}(\mathcal{U}_R)$$

Remark 1.1. A sequence of functions $\{f_n\}$ converges in $\mathcal{A}$ to f iff $||f_n - f||_R \to 0$ for every norm $||\cdot||_R$.

The function f is said to be analytic at zero if there exists $R \in \Gamma$ such that $f \in \mathcal{A}(\mathcal{U}_R)$. The space of functions analytic at zero, $\mathcal{A}_0(K^n) \equiv \mathcal{A}_0(K^n, K)(\equiv \mathcal{A}_0)$ is provided with a topology of an inductive limit:

$$\mathcal{A}_0(K^n) = \lim_{R\to 0} \mathrm{ind}\, \mathcal{A}(\mathcal{U}_R)$$

Remark 1.2. According to the theory of inductive limits a sequence of functions $\{f_n\}$ converges in the space $\mathcal{A}_0$ if all these functions belong to the same space $\mathcal{A}_R$ for some R and this sequence converges in this Banach space.

The space $\mathcal{A}_0$ is a complete locally convex space.

Remark 1.3. If the reader is not interested in topological constructions, he can consider only convergence of sequences of functions.

Proposition 1.1. *Differential operators* $\frac{\partial}{\partial x_j} : \mathcal{A}_\delta \to \mathcal{A}_\delta, j = 1, ..., n,$ *are continuous.*

Proof. We have

$$\frac{\partial}{\partial x_j} f(x) = \sum_{\alpha} \alpha_j f_\alpha x_1^{\alpha_1} ... x_j^{alpha_j - 1} ... x_n^{\alpha_n},$$

then

$$|\alpha_j|_K |f_\alpha|_K \delta^{|\alpha|-1} \leq (1/\delta)\delta^{|\alpha|} |f_\alpha|_K \to 0, |\alpha| \to \infty.$$

(we have used the estimate $|m|_K \leq 1$ for natural m).

Hence, $\frac{\partial}{\partial x_j} f(x) \in \mathcal{A}_\delta$. Further,

$$||\frac{\partial}{\partial x_j} f||_\delta \leq (1/\delta)||f||_\delta. \tag{1.2}$$

Using the definitions of the topologies in the spaces $\mathcal{A}$ and A_0, we get that the differential operators $\frac{\partial}{\partial x_j}$ are continuous in these spaces. We recall that if a linear operator is continuous in every Banach space of an inductive limit then it is continuous.

Proposition 1.2. *Spaces $\mathcal{A}_\delta$ are non-Archimedean Banach algebras.*

Proof. Let us restrict to the case $n = 1$. Let functions f, g belong to the space $\mathcal{A}_\delta$. Then

$$(\delta^m/|m!|_K)|(fg)^{(m)}(0)|_K \leq$$

$$\max_{0\leq k\leq m}\Big((\delta^k/|k!|_K)|f^{(k)}(0)|_K(\delta^{m-k}/|(m-k)!|_K)|g^{(m-k)}(0)|_K\Big).$$

We choose such N that if $s \geq N$ then

$$(\delta^s/|s!|_K)|f^{(s)}(0)|_K, (\delta^s/|s!|_K)|f^{(s)}(0)|_K \leq \epsilon/\max(||f||_\delta, ||g||_\delta).$$

Let $m \geq 2N$. If $k \geq N$ then

$$(\delta^m/|m!|_K)|(fg)^{(m)}(0)|_K \leq \epsilon||g||_\delta/\max(||f||_\delta, ||g||_\delta) \leq \epsilon.$$

If $k < N$ then $m - k \geq N$ and

$$(\delta^m/|m!|_K)|(fg)^{(m)}(0)|_K \leq \epsilon||f||_\delta/\max(||f||_\delta, ||g||_\delta) \leq \epsilon.$$

Thus, $(\delta^m/|m!|_K)|(fg)^{(m)}(0)|_K \to 0, m \to \infty$ and

$$||\, f\, g\, ||_\delta \;\leq\; ||f||_delta\, ||g||_\delta. \tag{1.3}$$

As a consequence of this proposition we get that the space $\mathcal{A}$ is a non-Archimedean Frechet algebra (because (1.3) is valid for every norm defining the topology in $\mathcal{A}$) and the space A_0 is a non-Archimedean locally convex algebra (because in an inductive limit inequalities (1.3) imply the continuity of the multiplication operation).

2. Analytic Distributions, Gauss and Feynman Distributions in a Finite-Dimensional Case

We choose the spaces of analytic functions $\mathcal{A}(K^n)$ and $\mathcal{A}_0(K^n)$ as the spaces of test functions and the spaces of K-linear continuous functionals $\mathcal{A}'(K^n)$ and $\mathcal{A}_0'(K^n)$ as the spaces of distributions (generalized functions). These spaces of distributions admit of an explicit description in the form of the spaces of infinite-order differential operators with coefficients from the field K.

For an infinite-order differential operator

$$P = \sum_{\alpha} \mathrm{P}_\alpha \delta^{(\alpha)}(x),$$

where $\mathrm{P}_\alpha \in K$, $\delta(x)$ is Dirac's function, we set

$$||\mathrm{P}||_\rho = \sup |\mathrm{P}_\alpha|_K |\alpha!|_K \rho^{-|\alpha|}.$$

We introduce spaces $\mathcal{D}_\rho = \{\mathrm{P} : \ ||\mathrm{P}||_\rho < \infty\}$. As usual the generalized derivative of Dirac's function is defined with the aid of the equality: $(\delta^{(\alpha)}(x), f(x)) = (-1)^\alpha f^{(\alpha)}(0)$ and a differential operator can be realized as a functional : $(P, f) = \sum_\alpha P_\alpha f^{(\alpha)}(0)$. The following theorem describe the spaces of distributions (continuous linear functionals) on the language of differential operators of an infinite order.

Theorem 2.1. $\mathcal{A}' = \cap \mathcal{D}_\rho$*and* $\mathcal{A}_0' = \cup \mathcal{D}_\rho$.

Thus a differential operator of an infinite order P is a distribution on the space of entire analytic functions iff $||P||_\rho < \infty$ for all $\rho > 0$. Further, a differential operator of an infinite order P is a distribution on the space of the functions analytic at the zero iff $||P||_\rho < \infty$ for some $\rho > 0$. The proof of the theorem is based on the following proposition.

Proposition 2.1. *The monomials* $e_\alpha = x^\alpha$ *form a basis in the locally convex spaces* $\mathcal{A}$ *and* $\mathcal{A}_0$ *and* $||e_\alpha||_R = R^{|\alpha|}$.

Thus the power expansion of every entire analytic function converges in the Frechet space $\mathcal{A}$ and this expansion converges in the non-Archimedean locally convex space $\mathcal{A}_0$ for every function analytic at the zero. The proof of this proposition is the direct estimate based on the definition of norms $||\cdot||_R$. Then Theorem 5.1 , Ch. 1, is applied. To finish the proof for the space $\mathcal{A}_0$ we must use the basic property of convergence sequences in inductive limits (see Remark 1.3).

In particular, this proposition generates representations of the spaces of test functions $\mathcal{A}$ and $\mathcal{A}_0$ as the spaces of K-sequences. Sometimes such representation will be very useful in computations. Theorem 2.1 generates such sequence-representation of the spaces of distributions $\mathcal{A}'$ and $\mathcal{A}'_0$. If $P=(P_\alpha)$ and $f=(f_\alpha)$ then $P(f)=(P,F)=\sum_\alpha P_\alpha f_\alpha \alpha!$.

Let us introduce topologies in the spaces of distributions using their structure proposed by Theorem 2.1: $\mathcal{A}'=\lim_{\rho\to\infty}\operatorname{ind}\mathcal{D}_\rho$, $\mathcal{A}'_0=\lim_{\rho\to 0}\operatorname{proj}\mathcal{D}_\rho$.

Proposition 2.2. *The monomials $e'_\alpha=(-1)^\alpha\delta^{(\alpha)}(x)$ form a basis in the space $\mathcal{A}'$ provided with the inductive topology and in the space $\mathcal{A}'_0$ provided with the projective topology.*

Note for further considerations (Ch. 3) that $(e'_\alpha,e_\alpha)=\alpha!$.

You could find more extended proofs on the properties of the spaces $\mathcal{A},\mathcal{A}_0,\mathcal{A}',\mathcal{A}'_0$ in the paper [57]. In particular, it was proved in [57] that the spaces of test functions $\mathcal{A}$ and $\mathcal{A}_0$ are reflexive : $\mathcal{A}''=\mathcal{A}$ and $\mathcal{A}''_0=\mathcal{A}''_0$.

To use analytic functions is the simplest way to a non-Archimedean generalization of the theory of distributions. This analytic theory is not so rich as possible non-analytic generalizations, see, for example, [106] , but it is sufficient to our applications to the non-Archimedean mathematical physics.

As usual, it is convenient to use the symbol of an integral to denote the effect produced by a distribution on the test function. The following symbols are convenient for us

$$J(f)=(J,f)=\int_{K^n} J(\mathrm{d}x)f(x),\ J\in\mathcal{A}'_0, f\in\mathcal{A}_0,$$

$$\mu(f)=(f,\mu)=\int_{K^n} f(x)\mu(\mathrm{d}x),\ \mu\in\mathcal{A}', f\in\mathcal{A}.$$

Thus, we write distributions of the class $\mathcal{A}_0'$ on the left side and distributions of the class $\mathcal{A}'$ on the right side to a test function.

A product of two distributions J_1 and $J_2 \in \mathcal{A}_0'$ is defined by

$$\int\limits_{K^{n+m}} J_1 \otimes J_2(\mathrm{d}u\mathrm{d}v) f(u,v) = \int\limits_{K^n} J_1(\mathrm{d}u) \int\limits_{K^m} J_2(\mathrm{d}v) f(u,v).$$

Simple (but sufficiently long) computations with power series show that $J_1 \otimes J_2$ is a well defined continuous K-linear functional on the space $\mathcal{A}_0$. Thus, it belongs to the space $\mathcal{A}_0'$. All estimates in these computations are based on the strong triangle inequality. So the operation of the product of distributions is well defined in the space $\mathcal{A}_0$.' A convolution of two distributions $J_1, J_2 \in \mathcal{A}_0'$ is defined with the aid of the equality :

$$\int\limits_{K^n} J_1 * J_2(\mathrm{d}v) f(v) = \int\limits_{K^{2n}} J_1 \otimes J_2(\mathrm{d}v_1\mathrm{d}v_2) f(v_1 + v_2).$$

It is a well defined operation on the space of distributions A_0'.

We define the operations of the product and the convolution on the space of distributions $\mathcal{A}$ in the same way. These operations are also well defined. The proofs of all these propositions about the properties of $\otimes$ and $*$ you could find in [69].

As usual, we can define the operation of the differentiation in the spaces of distributions $\mathcal{A}_0'$ and $\mathcal{A}'$: $(\frac{\partial J}{\partial y_j}, \phi) = -(J, \frac{\partial \phi}{\partial y_j}), J \in \mathcal{A}_0', \phi \in \mathcal{A}_0$, and in the same way for $\mu \in \mathcal{A}', \phi \in \mathcal{A}$. This operation is well defined as the operators $\frac{\partial}{\partial y_j}, j = 1, ..., n$, are continuous in locally convex spaces $\mathcal{A}_0$ and $\mathcal{A}$ (thus, adjoint operators are well defined).

As usual, we can also define the operation of the multiplication to a test function in the spaces of distributions :$(g\ J, \phi) = (J,\ g\phi), J \in \mathcal{A}_0', g, \phi \in \mathcal{A}_0$, and in the same way in the case $\mu \in \mathcal{A}', g, \phi \in \mathcal{A}$. This operation is well defined as the spaces of test functions are locally convex algebras.

We introduce an inner product on K^n, it is $(x,y) = \sum_{j=1}^n x_j y_j$.

Definition 2.1. *The Laplace transform (two-sided) of the distribution $g \in \mathcal{A}_0'$ is a function $L(g)(y) = (g, \exp\{(y, \cdot)\})$.*

As we have already discussed in Ch.1 the exponential function is analytic in some neighborhood of zero ;this neighbourhood depends on K. Hence the function $\exp\{(y,\cdot)\})$ is analytic in a neighbourhood of zero for every $y \in K^n$. Thus, it belongs to $\mathcal{A}_0$ for every $y \in K^n$ and the Laplace transform is well defined. This is why we consider the space of test functions $\mathcal{A}_0$ as the basis of the non-Archimedean Laplace transform.

Theorem 2.2. *The Laplace transformation $L : \mathcal{A}_0' \to \mathcal{A}$ is an isomorphism.*

We shall prove the same result directly in infinite dimensional case in Ch. 3. A full proof of this theorem in finite dimensional case can be found in [72].

Thus we have a non-Archimedean Laplace calculus

$$\mathcal{A}_0' \xrightarrow{L} \mathcal{A}, \qquad \mathcal{A}_0 \xleftarrow{L'} \mathcal{A}'. \tag{2.1}$$

The Laplace transformation has all standard properties of the usual Laplace transformation. Suppose that $J, J_1, J_2 \in \mathcal{A}_0'$ and $\mu, \mu_1, \mu_2 \in \mathcal{A}'$. Then:

1)$L(J_1 * J_2) = L(J_1)\ L(J_2)$ and $L'(\mu_1 * \mu_2) = L'(\mu_1)\ L'(\mu_2)$;

2)$L(y_j J) = \partial L(J)/\partial x_j$ and $L'(x_j\mu) = \partial L'(\mu)/\partial y_j$;

3)$L(\partial J/\partial y_j) = -x_j L(J)$ and $L'(\partial\mu/\partial x_j) = -y_j L'(\mu)$. These properties are the direct consequence of the definition.

By the definition of a conjugate operator, we have *Parseval's equality*

$$\int_{K^n} L(g)(y)\mu(dy) = \int_{K^n} g(\mathrm{d}x)L'(\mu)(x). \tag{2.2}$$

Definition 2.2. *The Gauss distribution on K^n (with the mean value $a \in K^n$ and the covariance matrix B) is $\gamma_{a,B} \in \mathcal{A}'$ with the Laplace transform $L'(\gamma_{a,B})(x) = \exp\{\frac{1}{2}(Bx, x) + (a, x)\}$.*

If we consider $\mathbb{R}$ instead of K, then we get ordinary Gaussian distribution for the matrix $B > 0$.

To denote the integral with respect to the Gaussian distribution, we use the symbol

$$\int_{K^n} \varphi(x) \exp\Big\{ -\frac{1}{2}(B^{-1}(x-a),(x-a))\Big\}\,\mathrm{d}x.$$

The quadratic exponent multiplied to $\mathrm{d}x$ is only the symbol to denote the Gaussian distribution $\gamma_{a,B}$. But this symbol is sufficiently convenient in computations. Formally we can work with quadratic exponent as with usual density with respect to $\mathrm{d}x$, but we need to apply the Laplace transform to justify such computations because at the moment the symbol $\mathrm{d}x$ is not defined. This symbolic expression for the Gaussian distribution $\gamma_{a,B}$ contains a convention on the unit normalization of the integral

$$\int_{K^n} \exp\Big\{ -\frac{1}{2}(B^{-1}(x-a),(x-a))\Big\}\,\mathrm{d}x.$$

This normalization differs from the standard real normalization constant $\sqrt{2\pi \det B}$. But we cannot investigate the problem of the normalization on the basis of our definition of the Gaussian integral as the quadratic density is only convenient symbol in our computations. The separate question is a definition of a non-Archimedean analogue of π. There were attempts to define a non-Archimedean π, but these definitions were not connected in any way to our Gaussian integral.

By virtue of Parseval's equality (2.2) we have

$$\int_{K^n} L(g)(x) \exp\Big\{ -\frac{1}{2}(B^{-1}(x-a),(x-a))\Big\}\,\mathrm{d}x =$$

$$\int_{K^n} g(\mathrm{d}x) \exp\Big\{\frac{1}{2}(Bx,x)+(a,x)\Big\}. \tag{2.3}$$

Thus, it is possible to compute every Gaussian integral, using only the Laplace transform.

Note also the formula for integration by parts for the Gaussian distribution: if $\varphi \in \mathcal{A}$, $a \in K^n$, then

$$\int_{K^n} \varphi(x)(a,x) \exp\Big\{ -\frac{1}{2}(B^{-1}x,x)\Big\}\,\mathrm{d}x =$$

$$\int_{K^n} \Big(Ba, \frac{\partial}{\partial x}\Big)(\varphi)(x) \exp\Big\{ -\frac{1}{2}(B^{-1}x, x)\Big\}\, \mathrm{d}x.$$

We propose the proof of this formula to demonstrate the Laplace transform technic for Gaussian integrals. Using (2.3) and the properties of L, we get

$$I = \int_{K^n} \varphi(x)(a, x) \exp\Big\{ -\frac{1}{2}(B^{-1}x, x)\Big\}\, \mathrm{d}x =$$

$$\int_{K^n} L^{-1}(\varphi) * \rho_a(\mathrm{d}y) \exp\Big\{\frac{1}{2}(By, y)\Big\},$$

where $\rho_a = (\operatorname{grad}\delta, a) = \sum_{j=1}^n a_j \frac{\partial \delta(x)}{\partial x_j}$. Hence

$$I = \int_{K^n} L^{-1}(\varphi) \otimes \rho_a(\mathrm{d}y_1 \mathrm{d}y_2) \exp\Big\{\frac{1}{2}(By_1 + y_2, y_1 + y_2)\Big\} =$$

$$\int_{K^n} L^{-1}(\varphi)(\mathrm{d}y)(Ba, y) \exp\Big\{\frac{1}{2}(By, y)\Big\}.$$

Further, we see that $L((Ba, \cdot)L^{-1}(\varphi)) = \Big(Ba, \operatorname{grad}\varphi\Big)$. To finish the proof we apply (2.3) again.

All the preceding considerations can be extended to the case of analytic functions $f: K^n \to Z, Z = K(\sqrt{\tau})$. Here we have the spaces $\mathcal{A}(K^n, Z)$ of entire functions, the space $\mathcal{A}_0(K^n, Z)$ of functions analytic at zero and the spaces of distributions $\mathcal{A}'(K^n, Z)$ and $\mathcal{A}'_0(K^n, Z)$. In the framework of this theory of distributions we introduce Gaussian distributions in which $a \in Z$ and the elements of the matrix B also belong to Z.

For the Z-value theory we also introduce the *Fourier transform*:

$$\mathcal{F}(g)(y) = \int_{K^n} g(\mathrm{d}x) \exp\{\sqrt{\tau}(y, x)\}.$$

Theorem 2.3. *The Fourier transformation $\mathcal{F}: \mathcal{A}'_0(K^n, Z) \to \mathcal{A}(K^n, Z)$ is an isomorphism.*

The Fourier transformation has the usual properties. The only exchange in the standard formula is $i \to \sqrt{\tau}$. If $K = Q_p$ and $p =$, then the quadratic extension $Z = Q_p(i)$, $i = \sqrt{-1}$, can be considered. In this case the Fourier transform is defined by the same formula as in the real case and all formula coincide with the complex results.

Example 2.1 (*the canonical Gaussian distribution* $\nu(\mathrm{d}x) = \mathrm{e}^{-x^2}\mathrm{d}x$),

$$M_n = \int\limits_K x^n \mathrm{e}^{-|x|^2}\,\mathrm{d}x = \frac{\partial^n}{\partial y^n}\mathrm{e}^{y^2/4}\bigg|_{y=0}.$$

Consequently, $M_{2k+1} = 0$, $M_{2k} = (2k-1)!!/2^k$. Using proposition 2.1, we get a formula for the calculation of the Gaussian integral of any entire function $f(x) = \sum_{n=0}^{\infty} f_n x^n$:

$$\int\limits_K f(x)\mathrm{e}^{-|x|^2}\,\mathrm{d}x = \sum_{n=0}^{\infty} f_{2n}(2n-1)!!/2^n.$$

According to the Gaussian distribution, many functions "exotic" from the point of view of the theory of real functions, are integrable. Let $K = \mathbb{Q}_p$. Then the function $f(x) = \sum_{n=0}^{\infty} n! x^n$ is integrable and

$$\int\limits_{\mathbb{Q}_p} f(x)\,\mathrm{d}x = \sum_{n=0}^{\infty}[(2n)!](2n-1)!!/2^n.$$

Let us introduce *Hermitian polynomials* over the field K: $H_n(x) = (-1)^n \mathrm{e}^{x^2}\frac{\mathrm{d}^n}{\mathrm{d}x^n}\mathrm{e}^{-x^2}$ and, as usual, $\int_K H_n(x)H_m(x)\mathrm{e}^{-x^2}\,\mathrm{d}x = 0$, $n \neq m$, calculate $\int_K H_n^2(x)\mathrm{e}^{-x^2}\,\mathrm{d}x$. To do this, we take into account that the product of the test function $H_n \in \mathcal{A}$ by the distribution $\nu \in \mathcal{A}'$ is equal to the generalized derivative of ν: $(-1)^n\frac{\mathrm{d}^n}{\mathrm{d}x^n}(\nu)$,

$$\int\limits_K H_n^2(x)\mathrm{e}^{-x^2}\,\mathrm{d}x = (-1)^n\int\limits_K H_n(x)\frac{\mathrm{d}^n}{\mathrm{d}x^n}(\mathrm{e}^{-x^2}\mathrm{d}x) =$$

$$= \int\limits_K \frac{\mathrm{d}^n}{\mathrm{d}x^n}(H_n(x))\mathrm{e}^{-x^2}\,\mathrm{d}x = 2^n n! \int\limits_K \mathrm{e}^{-x^2}\,\mathrm{d}x = 2^n n!.$$

Example 2.2. (the canonical Gaussian distribution $\gamma(\mathrm{d}z, \mathrm{d}\bar{z}) = \mathrm{e}^{-z\bar{z}}\mathrm{d}z\mathrm{d}\bar{z}$). In this section we set $K = Z$ and denote by z and $\bar{z}$ independent variables on Z; $w = (z, \bar{z})$ is a variable on Z^2, $z\bar{z} = \frac{1}{2}(Bw, w)$, where B has the form $\begin{pmatrix} 0 & 1 \\ 1 & 0 \end{pmatrix}$. Consequently, $\gamma(\mathrm{d}z, \mathrm{d}\bar{z})$ is a Gaussian distribution on Z^2 with a zero mean and a covariance matrix B, and

$$\int\limits_{Z^2} z^n \bar{z}^m \mathrm{e}^{-z\bar{z}}\,\mathrm{d}z\mathrm{d}\bar{z} = \frac{\partial^{n+m}}{\partial \bar{z}^n \partial z^m} \mathrm{e}^{\bar{z}z}\bigg|_{\substack{z=0\\ \bar{z}=0}} = \delta_{nm} n!.$$

For instance, the function $f(z, \bar{z}) = \sum_{n=0}^{\infty} \frac{p^{p^n}}{n!} z^n \bar{z}^n$ belongs to the space $\mathcal{A}(Z^2, Z)$ and we have

$$\int\limits_{Z^2} f(z, \bar{z})\mathrm{e}^{-z\bar{z}}\,\mathrm{d}z\mathrm{d}\bar{z} = \sum_{n=0}^{\infty} p^{p^n}$$

for it. As has been pointed out, we consider here the variables z and $\bar{z}$ as two independent variables from Z^2, $w_1 = z$, and $w_2 = \bar{z}$.

3. Non-Archimedean Hilbert Space

The quantization over non-Archimedean number fields must be based on the non-Archimedean analog of a Hilbert space. In the mathematical literature [101], [106] the concept of a non-Archimedean Hilbert space adequate for physical applications has not yet been worked out. The concept of orthogonality in non-Archimedean spaces is based not on an inner product but on a norm.

Recall [102], [106] that the system of vectors $\{e_j\}_{j\in J}$ in the non-Archimedean normalized space E is said to be orthogonal if

$$\left\| \sum_{j\in S} x_j e_j \right\| = \max_{j\in S} |x_j|_K \|e_j\|$$

for every finite set $S \subset J$ and any $x_j \in K$. The orthogonal system $\{e_j\}_{j\in J}$ is called an orthogonal basis in E if $x = \sum x_j e_j$ for any vector $x \in E$. In this case the space E is said to be orthogonalizable. By virtue of *J.-P. Serre's orthogonalization theorem*, every discrete normalized non-Archimedean Banach space is orthogonalizable.

There is no canonical way of defining the inner product $(\cdot,\cdot)$ in the orthogonalizable space E. Supposing that the vectors $\{e_j\}$ must be orthogonal not only with respect to the norm but also with respect to the inner product, we have

$$(x, x) = \sum \lambda_j x_j^2,$$

where $\lambda_j = (e_j, e_j)$. This series converges if and only if $\lim_{j\to\infty} |x_j|_K \sqrt{|\lambda_j|_K} = 0$. But the Banach space E consists of those x for which $\lim_{j\to\infty} |x_j|_K ||e_j|| = 0$. If $||e_j||^2 \in \Gamma$, then we can take as λ_j any elements of the field K such that $|\lambda_j|_K = ||e_j||^2$. Now if $||e_j||^2 \notin \Gamma$, then it is impossible, in general, to find $\lambda_j \in K$. It is natural to include the numbers λ_j into the definition of a non-Archimedean Hilbert space. In what follows we shall consider only countable sets of indices.

For the sequence $\lambda = (\lambda_n) \in K^\infty$, $\lambda_n \neq 0$ we set

$$\mathcal{H}_\lambda = \{f = (f_n) : \text{ the series } \sum f_n^2 \lambda_n \text{ converges}\}.$$

By virtue of the non-Archimedean Cauchy criterion we have

$$\mathcal{H}_\lambda = \{f = (f_n) : \lim_{n\to\infty} |f_n|_K \sqrt{|\lambda_n|_K} = 0\}.$$

In the space $\mathcal{H}_\lambda$ we introduce a norm relative to which the base vectors $e_j = (e_j^i) = (\delta_j^i)$, are orthogonal, $||f||_\lambda = \max_n |f_n|_K \sqrt{|\lambda_n|_K}$. The space $\mathcal{H}_\lambda$ is a non-Archimedean Banach space. On the space $\mathcal{H}_\lambda$ we introduce an inner product $(\cdot,\cdot)$ consistent with the length $|f|_\lambda^2 = \sum f_n^2 \lambda_n$ setting $(f, g)_\lambda = \sum f_n g_n \lambda_n$. The inner product $(\cdot,\cdot) : \mathcal{H}_\lambda \times \mathcal{H}_\lambda \to K$ is continuous and we have the *Cauchy–Buniakovski inequality*:

$$|(f, g)_\lambda|_K \leq ||f|| \, ||g||. \tag{3.1}$$

Definition 3.1. *The triplet* $(\mathcal{H}_\lambda,\ (\cdot,\cdot)_\lambda,\ ||\cdot||_\lambda)$ *is called a coordinate Hilbert space [64], [72].*

An inner product on the non-Archimedean linear space E is an arbitrary nondegenerated symmetric bilinear form $(\cdot,\cdot):\ E\times E\to K$. It is evidently impossible to introduce an analog of the positive definiteness of a bilinear form. For instance, for the field of p-adic numbers any element $\gamma\in\mathbb{Q}_p$ can be represented as $\gamma=(x,x)_\lambda$, $x\in\mathcal{H}_\lambda$ (see [17]).

The triplets $(E_j,\ (\cdot,\cdot)_j,\ ||\cdot||_j)$, $j=1,2$, where E_j are non-Archimedean Banach spaces, $||\cdot||_j$ are norms and $(\cdot,\cdot)_j$ are inner products satisfying (3.1), are isomorphic if the spaces E_1 and E_2 are algebraically isomorphic and the algebraic isomorphism $I:\ E_1\to E_2$ is isometric and unitary, i.e., $||Ix||_2=||x||_1$, $(Ix,Iy)_2=(x,y)_1$.

Definition 3.2. *The triplet* $(E,\ (\cdot,\cdot),\ ||\cdot||)$ *is a non-Archimedean Hilbert space if it is isomorphic to the coordinate Hilbert space* $(\mathcal{H}_\lambda,$ $(\cdot,\cdot)_\lambda,\ ||\cdot||_\lambda)$ *for a certain* λ *[64], [72].*

The isomorphic relation divides the class of Hilbert spaces into equivalence classes. we shall define the equivalence class of Hilbert spaces by some coordinate representative $\mathcal{H}_\lambda$.

Example 3.1. Let $\lambda=(1)$ and $\mu=(2^n)$. The spaces $\mathcal{H}_\lambda$ and $\mathcal{H}_\mu$ belong to the same class of equivalence for the field $K=\mathbb{Q}_p$, $p\neq 2$, and to different classes for the field $K=\mathbb{Q}_2$.

Non-Archimedean Hilbert spaces over quadratic extensions $Z=K(\sqrt{\tau})$ can be introduced by analogy. For the sequence $\lambda=(\lambda_n)\in K^\infty$, $\lambda_n\neq 0$, we set

$$\mathcal{H}_\lambda=\{f=(f_n)\in Z^\infty:\ \text{the series}\sum|f_n|^2\lambda_n\ \text{converges in the field}\ K\}$$

$$=\{f=(f_n):\ \lim_{n\to\infty}|f_n|_K\sqrt{|\lambda_n|_K}=0\};$$

$$||f||_\lambda=\max_n|f_n|_K\sqrt{|\lambda_n|_K};$$

$$(f,g)=\sum f_n\bar{g}_n\lambda_n;\qquad |f|_\lambda^2=(f,f)_\lambda=\sum|f_n|^2\lambda_n\in K.$$

The triplet $(\mathcal{H}_\lambda,\ (\cdot,\cdot)_\lambda,\ ||\cdot||_\lambda)$ is a non-Archimedean complex coordinate Hilbert space. The non-Archimedean complex coordinate

Hilbert space $(E, (\cdot,\cdot), ||\cdot||)$ is defined as an isomorphic image of a coordinate Hilbert space.

A conjugate of a non-Archimedean Hilbert space does not coincide with it. A nested Hilbert space is a quadruple of injected spaces $V \subset E \subset E' \subset V'$, where V is a topological linear space densely injected into E.

We shall also use a non-Archimedean Hilbert space over the field of complex p-adic numbers C_p in some physical models. There is a possibility to compute all square roots $\sqrt{\lambda_j}$ and this is why we need not to consider these coefficients at all. Thus, we need only to define the standard sequence space :

$$\mathcal{H}(C_p) = \{f = (f_n), f_n \in C_p : \text{ the series } \sum f_n^2 \text{ converges}\}.$$

But there is a serious difference with the usual complex case , because there is not any involution on the field C_p. This is why the only possibility to define an inner product on $\mathcal{H}(C_p)$ is the following one $(f,g) = \sum f_n g_n$. It is the hard problem in our further physical considerations that this inner product is valued in C_p and not in Q_p.

4. The Space $L_2(K^n, \mathrm{e}^{-|x|^2}\mathrm{d}x)$ of Functions Square Summable with Respect to the Canonical Gaussian Distribution

On the space $\mathcal{A}$ of entire functions (K^n, Z) we consider a canonical Gaussian distribution

$$\nu(\mathrm{d}x) = \mathrm{e}^{-|x|^2}\mathrm{d}x, \qquad |x|^2 = \sum_{j=1}^{n} x_j^2.$$

Using the fact that the space $\mathcal{A}$ is an algebra, we introduce on $\mathcal{A}$ an inner product

$$(f,g) = \int_{K^n} f(x)\bar{g}(x)\mathrm{e}^{-|x|^2}\,\mathrm{d}x \tag{4.1}$$

and the square of the length of the function f

$$|f|^2 = \int\limits_{K^n} |f(x)|^2 e^{-|x|^2}\, dx \in K. \tag{4.2}$$

The algebra $\mathcal{A}$ is topological and, consequently, the inner product $(\cdot,\cdot)$ is continuous on $\mathcal{A}\times\mathcal{A}$. We use the symbols $H_\alpha(x)$ to denote the Hermitian polynomials corresponding to the Gaussian distribution ν,

$$H_{(\alpha_1,\dots,\alpha_n)}(x_1,\dots,x_n) = H_{\alpha_1}(x_1)\dots H_{\alpha_n}(x_n).$$

Proposition 4.1. *The Hermitian polynomials $H_\alpha(x)$ are orthogonal with respect to the inner product $(\cdot,\cdot)$, and*

$$\int\limits_{K^n} H_\alpha^2(x) e^{-|x|^2}\, dx = 2^{|\alpha|}\alpha!.$$

This proposition is a direct corollary of Example 1.1

Theorem 4.1 (the estimate of the growth of the Hermitian coefficients for entire functions). *Let $f \in \mathcal{A}$. Then, for the Hermitian coefficients $\tilde{f}_\alpha = (f, H_\alpha)/(H_\alpha, H_\alpha)$, the estimate*

$$|\tilde{f}_\alpha|_K \le \frac{||f||_R}{|\alpha!|_K (R|2|_K)^{|\alpha|}} \tag{4.3}$$

is valid for all $R \ge 1/\sqrt{|2|_K}$.

Proof. Let $n = 1$. Using Example 1.1, we have

$$\pi_n = (f, H_n) = \sum_{j=0}^{\infty} f_{n+2j}(2j+1)\dots(2j+n)(2j-1)!!2^{-j}.$$

Hence

$$|\pi_n|_K \le \sup_j (|f_{n+2j}|_K R^{n+2j}) \left(\frac{1}{|2|_K R^2}\right)^j \frac{1}{R^n}.$$

Theorem 4.2. *The Hermitian polynomials $H_\alpha(x)$ form a basis in the space $\mathcal{A}$ of entire functions.*

Proof. Let $n = 1$. We shall estimate the norms of the Hermitian polynomials in $\mathcal{A}$. We use the recurrence relation

$$H_{n+1}(x) = 2xH_n(x) - 2nH_{n-1}(x):$$

$$||H_{n+1}||_R \leq |2|_K R \max(||H_n||_R, ||H_{n-1}||_R) \leq (|2|_K R)^{n+1},$$
$$R \in \Gamma, \qquad R \geq 1.$$

Using estimate (4.3.), we get

$$|\tilde{f}_n|_K ||H_n||_R \leq ||f||_{R_1} R^n / R_1^n |n!|_K \leq ||f||_{R_1} (bR/R_1)^n \to 0$$

if $R_1 > bR$. Consequently, the Fourier series with respect to the Hermitian polynomials $\sum \tilde{f}_n H_n(x)$ converges in the space $\mathcal{A}$ for all $f \in \mathcal{A}$. we shall prove that it converges to the function f. It is sufficient to show that $f^{(n)}(0) = g^{(n)}(0)$, where $g(x)$ is the sum of the series. Direct calculations (collection of terms) show that

$$g(0) = \sum \tilde{f}_n H_n(0) = f(0)$$

The derivatives at zero can be considered by analogy. We have thus proved that

$$f(x) = \sum_\alpha \tilde{f}_\alpha H_\alpha(x) \tag{4.4}$$

for any $f \in \mathcal{A}$.

Proposition 4.2. *Let $f, g \in \mathcal{A}$. Then*

$$(f, g) = \sum_\alpha \tilde{f}_\alpha \bar{\tilde{g}}_\alpha 2^{|\alpha|}. \tag{4.5}$$

In order to prove (4.5.), it is sufficient to use Theorem 4.1 and Proposition 4.1.

In the space $\mathcal{A}$ we introduce a norm relative to which the Hermitian polynomials are orthogonal:

$$||f|| = \max_\alpha |f_\alpha|_K \sqrt{|\alpha!|_K |2|_K^{|\alpha|}}.$$

The completion of the space of entire functions with respect to this norm is called the space of functions square summable by the Gaussian distribution ν and is denoted by $L_2(K^n, \nu)$.

Proposition 4.3. $L_2(K^n, \nu) = \{f = \sum_\alpha \tilde{f}_\alpha H_\alpha :$ *the series* $|f|^2 = \sum |\tilde{f}_\alpha|^2 \alpha!\, 2^{|\alpha|}$ *converges*$\} = \{f = \sum \tilde{f}_\alpha H_\alpha :\ \lim_{|\alpha|\to\infty} |\tilde{f}_\alpha|_K |\alpha!|_K |2|_K^{|\alpha|} = 0\}$

The inner product (4.5) is continuous on $L_2(K^n, \nu)$ and the Cauchy–Buniakovski inequality (3.1) holds true. The triplet $(L_2(K^n, \nu), (\cdot,\cdot), ||\cdot||)$ is a non-Archimedean complex Hilbert space of the class $\mathcal{H}_{(\alpha!2^{|\alpha|})}$. The quadruple $\mathcal{A}(K^n, Z) \subset L_2(K^n, \nu) \subset L'_2(K^n, \nu) \subset \mathcal{A}'(K^n, \nu)$ is a nested Hilbert space. we shall use the notations (4.1) and (4.2.) for the inner product and square of length on $L_2(K^n, \nu)$. Note that

$$\int_{K^n} f(x)\bar{g}(x)\mathrm{e}^{-|x|^2}\,\mathrm{d}x = \lim_{N\to\infty} \int_{K^n} f_N(x) g_N(x)\mathrm{e}^{-|x|^2}\,\mathrm{d}x,$$

where

$$g_N(x) = \sum_{|\alpha|\le N} \tilde{g}_\alpha H_\alpha(x), \qquad f_N(x) = \sum_{|\alpha|\le N} \tilde{f}_\alpha H_\alpha(x).$$

Proposition 4.4. *The injections*

$$\mathcal{A}(U_{\rho_1}, Z) \subset L_2(K^n, \nu) \subset \mathcal{A}(U_{\rho_2}, Z)$$

are valid, where

$$\rho_1 > \sqrt{\frac{b}{|2|_K}}, \qquad \rho_2 = \frac{1}{\sqrt{b|2|_K}}.$$

Proof. (1) Let $f \in L_2$. we shall prove that (4.4.) converges in $\mathcal{A}(U_{\rho_2}, Z)$:

$$|\tilde{f}_n|_K ||H_n||_{\rho_2} \le (|\tilde{f}_n|_K \sqrt{|n!2^n|_K})|2|_K^{n/2} b^{n/2} \rho_2^n,$$

with $||f||_{\rho_2} \le \max |\tilde{f}_n|_K ||H_n||_{\rho_2} \le ||f||$.

(2) Let $f \in \mathcal{A}(U_{\rho_1,Z})$. We use the formula

$$\tilde{f}_n = \frac{1}{|H_n|^2}\sum_{j=n}^{\infty} f_j j(j-1)\ldots(j-n+1)\int_K x^{j-n}\mathrm{e}^{-x^2}\,\mathrm{d}x$$

to calculate the Hermitian coefficients. Repeating the calculations carried out in Lemma 4.1, we find that

$$|\tilde{f}_n|_K \le ||f||_{\rho_1}/(\rho_1|2|_K)^n|n!|_K.$$

Let us consider a function $g(x) = \sum \tilde{f}_n H_n(x)$. we shall show that $g \in L_2(K^n,\nu)$. Indeed,

$$|\tilde{f}_n|_K\sqrt{|n!2^n|_K} \le ||f||_{\rho_1}/(\rho_1^n\sqrt{|2|_K^{\frac{n}{2}}|n!|_K}) \le ||f||_{\rho_1}\left(\frac{1}{\rho_1}\sqrt{b/|2|_K}\right)^n.$$

Furthermore,

$$|\tilde{f}_n|_K||H_n||_R \le \left(\frac{bR}{\rho_1}\right)^n,$$

i.e., if $R < \rho_1/b$, then Fourier series (4.4.) converges in the space $\mathcal{A}(U_R, Z)$. By analogy with the proof of the theorem, we can show that $\sum \tilde{f}_n H_n(x) = f|_{U_R}$.

Remark 4.1. *Thus, in a non-Archimedean case all entire functions are square summable with respect to canonical Gaussian distribution and all square summable functions are analytical.*

5. The Space $L_2(K^n, \mathrm{d}x)$ of Functions Square Summable with Respect the Lebesgue Distribution

We introduce a functional space

$$U(K^n, Z) = \{f(x) = \varphi(x)\mathrm{e}^{-|x|^2} : \varphi(x) \in \mathcal{A}(K^n, Z)\}.$$

The topology in the space U is induced from the space $\mathcal{A}$ of entire functions by the isomorphism $I: U \to \mathcal{A}$, $I(f)(x) = f(x)\mathrm{e}^{|x|^2}$. The space U is a reflexive non-Archimedean Fréchet space. We choose this space as the space of test functions and U' as the space of generalized functions. The space of generalized functions U' is isomorphic to the space $\mathcal{A}'$, $I' : \mathcal{A}' \to U'$.

Definition 5.1. *The Lebesgue distribution on the non-Archimedean space* K^n *is a generalized function* $\mathrm{d}x = I'(\nu) \in U'(K^n, Z)$.

For the Lebesgue distribution $\mathrm{d}x$ to act on the test function $f \in U$, we use the integral notation

$$(\mathrm{d}x, f) = \int_{K^n} f(x)\,\mathrm{d}x.$$

Note the obvious properties of the integral with respect to the Lebesgue distribution $\mathrm{d}x$.

1. Linearity:

$$\int_{K^n} (\lambda f(x) + \mu g(x))\,\mathrm{d}x = \lambda \int_{K^n} f(x)\,\mathrm{d}x + \mu \int_{K^n} g(x)\,\mathrm{d}x.$$

2. The limiting process under the integral sign:

$$\lim_{m\to\infty} \int_{K^n} f_m(x)\,\mathrm{d}x = \int_{K^n} f(x)\,\mathrm{d}x$$

if $f_m \to f$.

3. Fubini's theorem:

$$\int_{K^n} f(x)\,\mathrm{d}x = \int_K (\ldots \int_K f(x_1, \ldots, x_n)\,\mathrm{d}x_1 \ldots)\,\mathrm{d}x_n.$$

In order to prove Fubini's theorem, it is sufficient to note that the Lebesgue distribution $\mathrm{d}x$ can be represented as the product of distributions, i.e., $\mathrm{d}x = \mathrm{d}x_1 \otimes \ldots \otimes \mathrm{d}x_n$.

We introduce a functional space

$$W(K^n, Z) = \{f(x) = \varphi(x)\mathrm{e}^{-|x|^2/2} : \ \varphi \in \mathcal{A}(K^n, Z)\}.$$

The topology in the space W is induced from $\mathcal{A}$ by means of the isomorphism $J: W \to \mathcal{A}$, $J(f) = f\mathrm{e}^{|x|^2/2}$. Let $f, g \in W$, and then $f, g \in U$.

4. The formula for integration by parts ($f, g \in W$):

$$\int_{K^n} \frac{\partial f}{\partial x_j}(x) g(x)\,\mathrm{d}x = -\int_{K^n} f(x)\frac{\partial g}{\partial x_j}(x)\,\mathrm{d}x.$$

Proof. Indeed,

$$\int_K \Big(\frac{\partial f}{\partial x}(x)g(x) + f(x)\frac{\partial g}{\partial x}(x)\Big)\,\mathrm{d}x =$$

$$= \int_K \Big(\frac{\partial J(f)}{\partial x}(x)J(g)(x) + J(f)(x)\frac{\partial J(g)}{\partial x}(x) - 2xJ(f)(x)J(g)(x)\Big)\nu(\mathrm{d}x) =$$

$$= (L^{-1}(J(f)) \otimes L^{-1}(J(g)(x,y), (x+y)\exp\{(x+y)^2/4\}) -$$

$$-2\Big(L^{-1}(J(f)) \otimes L^{-1}(J(g))(x,y), \frac{\partial}{\partial x}\exp\{(x+y)^2/4\}\Big) = 0.$$

We use ordinary formulas to introduce the inner product and the square of the length on the functional space W:

$$(f, g) = \int_{K^n} f(x)\bar{g}(x)\,\mathrm{d}x, \tag{5.1}$$

$$|f|^2 = \int_{K^n} |f(x)|^2 \mathrm{d}x. \tag{5.2}$$

The inner product $(\cdot,\cdot): W \times W \to Z$ is continuous (inequality (3.1)) and the Hermite functions $\varphi_\alpha(x) = H_\alpha(x)\mathrm{e}^{-|x|^2/2}$ are orthogonal to (5.1.). By virtue of Theorem 4.1, the Hermite functions form the Schauder basis in the space W, i.e., for any function $f \in W$ the series

$$f(x) = \sum \tilde{f}_\alpha \varphi_\alpha(x),$$

where

$$\tilde{f}_\alpha = \int_{K^n} f(x)\varphi_\alpha(x)\,\mathrm{d}x \Big/ \int_{K^n} \varphi_\alpha^2(x)\,\mathrm{d}x$$

are Hermitian coefficients, converges in the space W, whence follows a coordinate representation for (5.1) and (5.2)

$$(f, g) = \sum_\alpha \tilde{f}_\alpha \bar{\tilde{g}}_\alpha \alpha! 2^{|\alpha|}, \tag{5.3}$$

$$|f|^2 = \sum_\alpha |f_\alpha|^2 \alpha! 2^{|\alpha|}. \tag{5.4}$$

On the functional space W we introduce a norm relative to which the Hermite functions are orthogonal:

$$||f|| = \max_\alpha |f_\alpha|_K \sqrt{|\alpha!|_K |2|_K^{|\alpha|}}.$$

The completion of the space $W(K^n, Z)$ according to this norm is called the space of functions square summable by the Lebesgue distribution and is denoted by

$$L_2(K^n, \mathrm{d}x) =$$

$$\{f(x) = \sum_\alpha \tilde{f}_\alpha \varphi_\alpha(x) : \quad \text{the series} \quad \sum_\alpha |f_\alpha|^2 \alpha! 2^{|\alpha|} \quad \text{converges}\} =$$

$$= \{f(x) = \sum_\alpha \tilde{f}_\alpha \varphi_\alpha(x) : \quad \lim_{|\alpha|\to\infty} |\tilde{f}_\alpha|_K \sqrt{|\alpha!|_K |2|_K^{|\alpha|}} = 0\}.$$

we shall use notations (5.1) and (5.2) for the inner product (5.3) and the square of the length (5.4).

The isomorphism $J : W \to \mathcal{A}$ is continued to the isomorphism of the Hilbert space $J : L_2(K^n, \mathrm{d}x) \to L_2(K^n, \nu)$. In particular, the following nested Hilbert space appears:

$$W(K^n, Z) \subset L_2(K^n, \mathrm{d}x) \subset L_2'(K^n, \mathrm{d}x) \subset W'(K^n, Z).$$

Proposition 5.1. *The injection*

$$L_2(K^n, \mathrm{d}x) \subset \mathcal{A}(U_\delta, Z), \qquad \delta < \sqrt{|2|_K/b}.$$

is valid.

Proof. The function $e^{-|x|^2/2}$ belongs to the space $\mathcal{A}(U_\gamma, Z)$ for all $\gamma < \sqrt{|2|_K/b}$. By virtue of Proposition 4.3, we have

$$L_2(K^n, dx) \subset L_2(K^n, \nu)e^{-|x|^2/2} \subset \mathcal{A}(U_{1/\sqrt{b|2|_K}}, Z)\mathcal{A}(U_\gamma, Z).$$

Remark 5.1. *Thus, in a non-Archimedean case, all functions square summable by the Lebesgue distribution* dx *are analytic.*

We can introduce by analogy the space of K-valued functions square summable by the distributions of Gauss and Lebesgue.

6. The Space $F_2(Z^n, \gamma)$ of Analytic Functions Square Summable with Respect to the Canonical Complex Gaussian Distribution

On the space $\mathcal{A}(Z^{2n}, Z)$ of entire functions we consider a canonical Gaussian distribution $\gamma(dz, \bar{z}) = \exp\{-(z, \bar{z})\}dzd\bar{z}$ (in a multidimensional case it is defined in the same way as in a one-dimensional case, Example 1.2). We introduce an inner product and the square of length on the space $\mathcal{A}(Z^n, Z)$:

$$(f, g) = \int_{Z^{2n}} f(z)\overline{g(z)}e^{-(z,\bar{z})}\, dzd\bar{z}; \qquad |f|^2 = (f, f). \tag{6.1}$$

The monomials $z^\alpha = z_1^{\alpha_1} \dots z_n^{\alpha_n}$, $\alpha_j = 0, 1, \dots$, are orthogonal to (6.1.), with $|z^\alpha|^2 = \alpha!$. The representation in coordinates $(f, g) = \sum_{\alpha=0}^\infty f_\alpha \bar{g}_\alpha \alpha!$ is valid for the inner product (6.1). On the space $\mathcal{A}(Z^n, Z)$ we introduce a norm relative to which the monomials z^α : $||f|| = \max_\alpha |f_\alpha|_K \sqrt{|\alpha!|_K}$ are orthogonal. The completion of the space $\mathcal{A}(Z^n, Z)$ with respect to this norm is known as the space of Z-analytic functions which are square summable by the Gaussian distribution γ and is denoted $F_2(Z^n, \gamma)$.

Proposition 6.1. $F_2(Z^n, \gamma) = \{f(z) = \sum_\alpha f_\alpha z^\alpha$: *the series* $|f|^2 = \sum_\alpha |f_\alpha|^2 \alpha!$ *converges in* $K\} = \{f(z) : \lim_{|\alpha| \to \infty} |f_\alpha|_K \sqrt{|\alpha!|_K} = 0\}$.

Theorem 6.1. *The injection $F_2(Z^n, \gamma) \subset \mathcal{A}(U_\rho, Z)$, $\rho = 1/\sqrt{b}$, is valid.*

The quadruple

$$\mathcal{A}(Z^n, Z) \subset F_2(Z^n, \gamma) \subset F_2'(Z^n, \gamma) \subset \mathcal{A}'(Z^n, Z)$$

is a nested Hilbert space.

Theorem 6.2. *Let $\sqrt{2} \in Z$, and then the Hilbert spaces $L_2(K^n, dx)$ and $F_2(Z^n, \gamma)$ are isomorphic.*

The isomorphism $S: L_2(K^n, dx) \to F_2(Z^n, \gamma)$ is defined by the ordinary integral operator $S(\varphi)(z) = \int_{K^n} \varphi(x) \exp\{-\frac{1}{2}(|x|^2 - 2\sqrt{2}(x, z) + (z, z))\} \, dx$.

We can compare the spaces of analytic functions square summable by the Gaussian distribution γ in an Archimedean case $F_2(\mathbb{C}, \gamma)$ and in a non-Archimedean case $F_2(Z, \gamma)$. All polynomials with rational coefficients belong to both spaces. The function $f(z) = \sum_{n=1}^{\infty} n z^{n+1}$ does not belong to $F_2(\mathbb{C}, \gamma)$ but belongs to

$$F_2(Z, \gamma): \ |f|^2 = \int_{Z^2} |f(z)|^2 e^{-(z,z)} \, dz d\bar{z} = \sum_{n=1}^{\infty} n^2 (n+1)! = 2.$$

The function $f(z) = \sum_{n=0}^{\infty} \frac{z^n}{n!}$ does not belong to $F_2(\mathbb{Z}, \gamma)$ but belongs to $F_2(\mathbb{C}, \gamma): \ |f|^2 = e$. Thus a transition to a non-Archimedean case makes it possible to sum up the functions which cannot be summed up in the ordinary theory. On the other hand, in a non-Archimedean case many functions are no longer summable.

7. Unboundedness of the p-adic Gaussian Distribution

The non-Archimedean Gaussian distribution is well defined on the space of entire analytic functions. Is it possible to extend the Gaussian distribution to a measure and to define a Gaussian integral of a continuous function? The answer to this question turns out to be negative (in contrast to the real case). This section contains (fairly nontrivial) proof that the p-adic Gaussian distribution is unbounded. As usual, in p-adic theories the case $p = 2$ turns out to be "pathological". Our proof does not work in this case for a large class of Gaussian correlations. Thus, in this case the question of boundedness of the p-adic Gaussian distribution remains open.

Let λ be a measure on the ring of p-adic integers $\mathbb{Z}_p$, see Ch. 1. Then λ is defined with the aid of coefficients $\lambda_n = \lambda(C_x^n)$, where $C_x^n = (1/n!)x(x-1)...(x-n+1)$ are binomial polynomials. As a measure is a bounded $\mathbb{Q}_p$-linear functional on the space of continuous functions $C(\mathbb{Z}_p)$, a sequence λ_n is bounded. From the other hand, if we consider a $\mathbb{Q}_p$-linear functional λ on the space of polynomials such that the corresponding sequence $\{\lambda_n\}$ is bounded then this functional can be extended to a continuous $\mathbb{Q}_p$-linear functional on the space of continuous functions $C(\mathbb{Z}_p)$, a $\mathbb{Q}_p$-valued measure.

Let λ be a $\mathbb{Q}_p$-linear functional on the space of polynomials. We can define the formal power series $f(x) = f_\lambda(x)$:

$$f(x) = \sum_{n=0}^{\infty} \lambda_n x^n.$$

As usual, a formal power series is said to be *bounded* if the coefficients $\{\lambda_n\}$ are bounded. A formal power series $f(x) = f_\lambda(x)$ is bounded if and only if the functional λ is bounded. The correspondence (bounded formal power series $f(x) \to$ measures $\lambda = \lambda_f$) is one-to-one and invertible. Much more information in this direction you can find in [70].

Further, if

$$\lambda(x^n) = \int_{\mathbb{Z}_p} x^n d\lambda(x) = \alpha_n, \lambda(C_x^n) = \int_{\mathbb{Z}_p} C_x^n d\lambda(x) = \lambda_n,$$

then

$$g(z) = f(e^z - 1) = \sum_{n=0}^{\infty} \lambda_n (e^z - 1)^n = \sum_{n=0}^{\infty} \frac{\alpha_n}{n!} z^n.$$

Consequently,

$$f(x) = \sum_{n=0}^{\infty} \lambda_n x^n = \sum_{n=0}^{\infty} \frac{\alpha_n}{n!} (\log(1+x))^n,$$

where

$$\log(1+x) = \sum_{k=1}^{\infty} \frac{(-1)^{k-1}}{k} x^k.$$

Let us try to realize a p-adic Gaussian distribution $\lambda = \gamma_{0,b}$, $b \in \mathbb{Q}_p$ (one dimensional case) as a measure on $\mathbb{Z}_p$. The basis of our considerations are the momentum formulas

$$\alpha_{2k} = \frac{2k!}{k!}\left(\frac{b}{2}\right)^k, \quad \alpha_{2k+1} = 0, \ k = 0, 1, 2, ...$$

The formal power series corresponding to a Gaussian distribution has the form

$$f(x) = \sum_{n=0}^{\infty} \lambda_n x^n = \exp\{(b/2)(\log(1+x))^2\} = \sum_{n=0}^{\infty} \frac{1}{n!}\left(\frac{b}{2}\right)^n (log(1+x))^{2n}.$$

It will be proved below that the coefficients of the function $f(x)$ are unbounded.

Theorem 7.1. (Endo-Khrennikov)*Let $b/2 = p^a u$, where a is an integer and $|u|_p = 1$.*

1) If a is even, then for all primes p the coefficients $\{\lambda_n\}$ are unbounded.

2) If a is odd, then for all odd p the coefficients $\{\lambda_n\}$ are unbounded. For $l = 2np^{[a/2]+1}$, where $n = p^b$ (a power of p) , the equality

$$|\lambda_n|_p = p^{2n(1-\delta)+(n-1)/(p-1)}$$

is valid, where $\delta = 0$ if a is even , and $\delta = 1/2$ if a is odd.

The proof of this theorem will be based on a series of lemmas.

Lemma 7.1. *Let $n = a_0 + a_1 p + ... + a_t p^t$ be the p-adic decomposition of a positive integer n. Then the number $n!$ is divisible by the power p^g of p, where $g = (n-s)/(p-1)$ with $s = s_n = a_0 + ... + a_n$.*

It is well known result of the theory of numbers. See, for example, [17] , [106]. In particular, the important estimate (1.5) of Ch. 1 is based on Lemma 7.1.

Lemma 7.2. *Suppose that $x_1, ..., x_n$ are nonnegative real variables, and let A be a fixed positive number. Under the condition $x_1 + ... +$*

$x_n = A$ the product $x_1 \cdots x_n$ attains its maximal value $(A/n)^n$ when $x_1 = \ldots = x_n = A/n$.

Lemma 7.3. *Let $f(x) = (A/x)^x$, a function of a positive real variable. Then $f(x)$ attains its maximal value for $x = A/e$; the function $f(x)$ is monotonically increasing on $(0, A/e)$ and monotonically decreasing on $(A/e, \infty)$.*

Lemma 7.4. *Let $n = p^b$ be a power of p, and m a fixed integer with $n < m < np$. Assume that the value $1/|x_1 \cdots x_{2m}|_p$ attains its maximum for integer variables $x_1, \ldots, x_{2m}$ satisfying $x_1 + \cdots + x_{2m} = 2np^{c+1}$. Then it can be assumed that*

$$x_1 = x_2 = \ldots = x_{2m-u} = p^{c+1},$$

$$x_{2m-u+1} = x_{2m-u+2} = \ldots = x_{2m-1} = p^c,$$

$$x_{2m} = kp^c, 1 \leq k \leq p-1,$$

for some u.

Proof. By changing the order of variables if necessary, we can assume that $|x_1|_p \leq |x_2|_p \leq \ldots \leq |x_{2m}|_p$. We can also assume that all but the last of the x_i are powers of p. Indeed, if $x_i = kp^d$ for $k \geq 2$, then we set $y_i = p^d$ and $y_{2m} = x_{2m} + (k-1)p^d$. Then $|y_{2m}|_p \leq |x_{2m}|_p$, and hence

$$1/|x_1 x_2 \cdots x_i \cdots x_{2m}|_p \leq 1/|x_1 x_2 \cdots y_i \cdots y_{2m}|_p.$$

Thus, it can be assumed that $x_1 = p^d, \ldots, x_{2m-1} = p^e, x_{2m} = kp^f$ with $d \geq \ldots \geq e \geq f$ and $(k,p) = 1$. We show below that $d = c+1, f = c$, and $k \leq p-1$.

We show first that $f \geq c$. If $f < c$, then it follows from the condition $x_1 + \cdots\ x_{2m} = 2np^{c+1}$ that $d \geq c+1$. We get from the same condition that $f = e = \ldots$, and the sum $z = x_{2m-v+1} + x_{2m-v+2} + \cdots + x_{2m}$ must be divisible by p^{f+1} for some v $(\leq p)$. We get that $y_1 = p^{d-1}, \ldots, y_{v-1} = p^{d-1}, y_v = (p-v+1)p^{d-1}$. Then

$$y_1 + \cdots + y_v + x_2 + \cdots + x_{2m-v} + z = 2np^{c+1},$$

$$1/|x_1 x_2 \cdots x_{2m}|_p < 1/|y_1 \cdots y_v x_2 \cdots x_{2m-v} z|_p.$$

This contradicts the maximality of $1/|x_1x_2\cdots x_{2m}|_p$. Consequently, f must be $\geq c$.

We show next that $d \leq c+1$. If $d > c+1$, then it follows from the condition $x_1 + \cdots + x_{2m} = 2np^{c+1}$ that $f = c$. By the same arguments as above, the sum $z = x_{2m-v+1} + x_{2m-v+2} + \cdots + x_{2m}$ must be divisible by p^{c+1}. We set $y_1 = p^{d-1}, ..., y_{v-1} = p^{d-1}$, and $y_v = (p - v + 1)p^{d-1}$. Then also

$$1/|x_1x_2\cdots x_{2m}|_p < 1/|y_1\cdots y_v x_2\cdots x_{2m-v}z|_p,$$

i.e., we arrive at a contradiction. Consequently, d must be $\leq c+1$.

Finally, we show that $k \leq p-1$. If $k = qp+r$, then we set $y_{2m-1} = qp^{f+1}$ and $y_{2m} = (r+1)p^f$. As above,

$$1/|x_1x_2\cdots x_{2m}|_p < 1/|x_1x_2\cdots y_{2m-1}y_{2m}|_p.$$

This contradict the maximality. Thus, k must be $\leq p-1$.

Proof of Theorem 7.1. Let

$$f(x) = \sum_{n=0}^{\infty} \lambda_n x^n = \sum_{n=0}^{\infty} \frac{1}{n!}\left(\frac{b}{2}\right)^n (\log(1+x))^{2n}.$$

It must be shown that the coefficients $\{\lambda_l\}$ are unbounded ; for example, the sequence $\{|\lambda_{l_k}|_p\}_{k=1}^{\infty}$ contains the arbitrary large element $|\lambda_{l_s}|_p$. Recall that

$$\log(1+x) = \sum_{k=1}^{\infty} \frac{(-1)^{k-1}}{k} x^k.$$

Consequently, the coefficients λ_l of x^l are sums of the form

$$\sum_{m\leq l/2} \frac{(-1)^l}{m!}\left(\frac{b}{2}\right)^m \sum_{x_1,x_2,...,x_{2m}} \frac{1}{x_1x_2\cdots x_{2m}}, \tag{7.1}$$

where the summation is over all positive integers $x_1, ..., x_{2m}$ satisfying the condition $x_1 + x_2 + \cdots + x_{2m} = l$.

Consider the case $b/2 = p^a u$, where a is an integer and $|u|_p = 1$.

Let $a = 2(k+\delta)$ with integer k, i.e., $\delta = 0$ if a is even and $\delta = 1/2$ if a is odd. Also let $l = 2np^{k+1}$. It will be shown below that the values $|\lambda_l|_p$ become arbitrary large with increasing $n = p^b$ (a power of p). Actually, we get the estimate

$$|\lambda_l|_p = p^{2n(1+\delta)+(n-1)/(p-1)}, \tag{7.2}$$

which is equal $p^{2n+(n-1)/(p-1)}, a = 2k$, and $p^{n+(n-1)/(p-1)}, a = 2k+1$, for $n = p^b$(a power of p). We show that precisely one term in the sum (7.1) for the coefficient λ_l takes the maximal value (7.2), while all the other terms in this sum are strictly less than then the value (7.2). Then using Proposition 1.1, Ch. 1, we get the equality (7.2) for the coefficients λ_l.

Let $l = 2np^{k+1} = 2p^bp^{k+1} = 2p^{b+k+1}$ be fixed. Define

$$M(x_1, x_2, ..., x_{2m}) = |\frac{1}{m!}(\frac{b}{2})^m(1/x_1x_2...x_{2m})|_p$$

and

$$M(m) = \max\{M(x_1, x_2, ..., x_{2m}) : x_1 + x_2 + \cdots + x_{2m} = 2np^{k+1}\},$$

where $x_1, ..., x_{2m}$ are positive integer variable. It will be shown that only $M(n) = M(x_1, x_2, ..., x_{2m})(x_1 = ... = x_{2n} = p^{k+1})$ takes the maximal value (7.2) among all the terms in the coefficients λ_l.

It follows from Lemma 7.2 that $x_1 \cdots x_{2m} \le ((2n/2m)p^{k+1})^{2m}$. If $x_1 \cdots x_{2m} = p^c u$ with $(p, u) = 1$, then $|x_1 \cdots x_{2m}|_p = p^{-c}$. Since all the $x_1, ..., x_{2m}$ are positive integers, we get

$$M(x_1, x_2, ..., x_{2m}) = p^{(m-s)/(p-1)-am}|(1/x_1x_2...x_{2m})|_p =$$

$$p^{(m-s)/(p-1)-am}p^c \le p^{(m-s)/(p-1)-am}p^c u = p^{(m-s)/(p-1)-am}x_1x_2\cdots x_{2m} \le$$

$$\left(\frac{n}{m}p^{1-\delta+1/2(p-1)}\right)^{2m} p^{m-s/(p-1)}.$$

Consider the case $m = n$. If $(x_1, ..., x_{2n}) \ne (p^{k+1}, ..., p^{k+1})$, we have by Lemma 7.2 that $x_1 \cdots x_{2n} < p^{2n(k+1)}$. Consequently, from the preceding considerations

$$M(x_1, x_2, ..., x_{2n}) < p^{2n(1-\delta)+(n-1)/(p-1)}.$$

This shows that

(a) If $m = n$, then the maximum $M(n) = p^{2n(1-\delta)+(n-1)/(p-1)}$ is attained only in the case $x_1 = x_2 = ... = x_{2n} = p^{k+1}$.

Since $0 \leq \delta < 1$ and $ep^{-1} < p^{1-\delta+1/2(p-1)} < ep$, we get the following from Lemma 7.3:

(b) If $m > np$, then $M(m) < M(np)$.

(c) If $m < n/p$, then $M(m) < M(n/p)$.

Furthermore,

$$M(np) = (p^{-\delta+1/2(p-1)})^{2np} p^{-1/(p-1)} < M(n).$$

We now show that:

(c) If $n < m < np$, then $M(m) < M(n)$.

(d) If $n/p < m < n$, then $M(m) < M(n/p)$ when $\delta > (2p-3)/2(p-1)$ and $M(m) < M(n)$ when $\delta < (2p-3)/2(p-1)$.

(e) $M(n/p) < M(n)$ if and only if $\delta < (2p-3)/2(p-1)$.

Of course, all our arguments are carried out in the case $\delta = 0$ (even a) or $\delta = 1/2$(odd a). Thus, with the exception of the case $p = 2$ and a is odd, it will be shown that $M(n)$ is the maximal value.

We proceed to the proof of (c). Let $n < m < np$. If $M(x_1, ..., x_{2m})$ attains the maximum $M(m)$, then we can assume according to Lemma 7.4 that

$$x_1 = x_2 = ... = x_{2m-u} = p^{k+1}, x_{2m-u+1} = ... = x_{2m-1} = p^k,$$

$$x_{2m} = ip^k, 1 \leq i \leq p-1.$$

Since $x_1 + x_2 + ... + x_{2m} = 2np^{k+1}$, we get

$$(2m-u)p^{k+1} + (u-1)p^k + ip^k = 2np^{k+1},$$

$$2(m-n)p = u(p-1) - (i-1). \tag{7.3}$$

We next observe that

$$M(m) = M(x_1, ..., x_{2m}) = p^{2(m-s)/(p-1)-am+(2m-u)(k+1)+uk} =$$

$$p^{(m-s)/(p-1)+2m(1-\delta)-u}.$$

Since $M(n) = p^{(n-1)/(p-1)+2n(1-\delta)}$, it follows that

$$M(n)/M(m) = p^{2(n-m)(1-\delta)+(n-m)/(p-1)+u+(s-1)/(p-1)}.$$

Using (7.3), we get that the exponent on the right-hand side of the preceding formula has the form

$$2(n-m)\left\{\frac{1}{2(p-1)} + (1-\delta)\right\} + \frac{1}{p-1}\{2(m-n)p + (i-1)\} + \frac{s-1}{p-1}$$

$$= 2(m-n)\left\{\frac{1}{2(p-1)} + \delta\right\} + \frac{i-1}{p-1} + \frac{s-1}{p-1} > 0.$$

Thus, $M(n) > M(m)$.

Let us proceed to proof of (d). Here the situation is similar to (c). If $n/p < m < n$ and $M(x_1, ..., x_{2m})$ attains the maximum $M(m)$, then by Lemma 7.4 we can assume that

$$x_1 = x_2 = ... = x_{2m-u} = p^{k+2},\ x_{2m-u+1} = ... = x_{2m-1} = p^{k+1},$$

$$x_{2m} = ip^{k+1}, 1 \le i \le p-1.$$

Further,

$$(2m-u)p^{k+2} + (u-1)p^{k+1} + ip^{k+1} = 2np^{k+1},$$

$$2(pm-u) = (p-1)u - (i-1).$$

Since

$$M(m) = p^{(m-s)/(p-1)+2m(2-\delta)-u}$$

and

$$M(n/p) = p^{(n/p-1)/(p-1)+2n(2-\delta)/p},$$

it follows that

$$M(n/p)/M(m) = p^{(s-1)/(p-1)+(n-pm)/(p(p-1))+2((n-pm)/p)(2-\delta)+u}.$$

It is not hard to see that the exponent on the right-hand side of the preceding formula has the form

$$\frac{n-mp}{p(p-1)} + \frac{s-1}{p-1} + \frac{2(n-mp)}{p}(2-\delta) + \frac{2(mp-n)}{p-1} + \frac{i-1}{p-1}$$

$$= \frac{mp-n}{p(p-1)}[-1-(p-1)(2-\delta)+2p] + \frac{s-1}{p-1} + \frac{i-1}{p-1}.$$

This shows that $M(n/p) > M(m)$ if $\delta > (2p-3)/2(p-1)$.

Next, let $\delta < (2p-3)/2(p-1)$. Then

$$\frac{M(n)}{M(m)} = \frac{p^{(n-1)/(p-1)+2n(1-\delta)}}{p^{(m-s)/(p-1)+2m(2-\delta)-u}},$$

i.e., the exponent has the form

$$\frac{n-m}{p-1} + \frac{s-1}{p-1} + 2(n-m)(1-\delta) - 2m + 2m + 2\frac{m-n}{p-1} + \frac{i-1}{p-1}.$$

Consequently, in this case $M(m) < M(n)$.

Finally, we prove (e). It is clear that

$$\frac{M(n)}{M(n/p)} = p^{(n/p)\{2(1-\delta)(p-1)-1\}}.$$

This implies that $M(n) > M(n/p)$ if and only if $2(1-\delta)(p-1) > 1$, i.e., $\delta < (2p-3)/2(p-1)$.

We considered Gaussian distributions with $b \in \mathbb{Q}_p$. There is no problem to generalize our considerations to the case $b \in \mathbb{C}_p$. The following result is valid:

Theorem 7.2. *Let $|b/2|_p = p^{-a}$, where $a = 2(k+\delta), 0 \leq \delta < 1$.*

1) If $\delta < (2p-3)/2(p-1)$, then $M(n) = p^{2n(1-\delta)+(n-1)/(p-1)}$ is the unique maximal value.

2) If $\delta > (2p-3)/2(p-1)$, then $M(n/p) = p^{2n(2-\delta)/p+(n-p)/p(p-1)}$ is the unique maximal value.

The main problem is the following:

Is a p-adic Gaussian integral bounded or unbounded in the case when b belongs to the field $\mathbb{C}_p$ and $\delta = (2p-3)/2(p-1)$? (In particular, in the case when $p = 2$ and a is odd.)

Recently, author has received a letter of M.Endo with some intuitive considerations on boundedness of a 2-adic Gaussian distributions for some b. But these were only the ideas. And there is no rigorous mathematical results in this direction.

8. The Volkenborn (Uniform) Distribution

The Volkenborn "measure" was already studied in Ch. 1. But it was not a well defined mathematical object, as this "measure" was not really a measure. It is unbounded on the algebra of clopen sets. Now this "measure" will be realized as a distribution (generalized function), an element of $\mathcal{A}'(\mathbb{Q}_p)$. Using the definition of Ch. 1, we get that

$$u(a) = \int_{\mathbb{Z}_p} \mathrm{e}^{ax}\mathrm{d}x = a/(\mathrm{e}^a - 1)] = \sum_{n=0}^{\infty} B_n a^n/n!.$$

This function belongs to the space $\mathcal{A}_0$. And we introduce the Volkenborn distribution as the element of $\mathcal{A}'$ with the Laplace transform $L'(m)(a) = u(a)$.

At the moment, it is not clear for me a connection between the Volkenborn distribution and the Lebesgue distribution. Intuitive considerations show that these objects must coincide (and we use the same symbol $\mathrm{d}x$). But it is an open problem to realize this on a mathematical level. The main problem is different classes of functions on which these distributions are defined.

Chapter III

The Gauss and Feynman Distributions on Infinite-Dimensional Spaces over Non-Archimedean Fields

By analogy with a finite-dimensional case (Ch. 2) the Gauss and Feynman integrals are defined on the basis of the distribution theory. A variant of the distribution theory is suggested in [69] where the spaces of analytic functions on infinite-dimensional non-Archimedean spaces are used as the spaces of test functions (the functions of an infinite-dimensional argument are often called functionals, but we shall not use this term here). This theory of non-Archimedean spaces is a natural generalization, to the non-Archimedean case, of the theory of analytic infinite-dimensional distributions [38], [43–45], [47], [48] over the field of complex numbers and the supercommutative Banach superalgebras [49], [50], [52], [55], [56].

Other (nonanalytic) theories of distributions on infinite-dimensional spaces have also been constructed for the field of complex numbers. These are distributions on a space with a fixed Gaussian measure, as well as distributions and generalized measures corresponding to the theory of Fomin's differentiable measures. For these theories see [7], [34–36] (Gauss' theory) and [11], [115], [134] (a theory invariant relative to the choice of a measure).

It seems to me that similar theories can also be suggested in a non-Archimedean case. A particular case (the functionals of the p-

adic white noise) is considered in [64], see Ch. 6.

1. Continuous Multilinear Forms

For the non-Archimedean Banach space E we set

$$\Gamma_E = \{r \in \mathbb{R}_+ : \ r = ||x||, \ x \in E\} \equiv ||E||.$$

Lemma 1.1. $\Gamma_{A(U_R,K)} = \Gamma$.

Proposition 1.1. *Let E be a non-Archimedean Banach space for which $\Gamma_E = \Gamma$. Then*

$$\sup_{||x_j|| \leq 1} |b(x_1, \ldots, x_n)|_K = \sup_{||x_j|| \neq 0} |b(x_1, \ldots, x_n)|_K = /||x_1|| \ldots ||x_n||.$$

Proof. 1. Indeed, $||b|| = \sup_{||x_j|| \leq 1} |b(x_1, \ldots, x_n)|_K \leq |||b||| = \sup_{x_j \neq 0} |b(x_1, \ldots, x_n)|_K / ||x_1|| \ldots ||x_n||$.

2. Using the fact that $\Gamma_E = \Gamma$, we find that there is an element $a_j : \ |a_j|_K = ||x_j||$. Consequently,

$$|||b||| = \sup_{x_j \neq 0} |b(x_1 a_1^{-1}, \ldots, x_n a_n^{-1})|_K \leq ||b||.$$

By virtue of Lemma 1.1, Proposition 1.1 is applicable to continuous multilinear forms on the space of entire functions A (every continuous multilinear form is continuous by a certain norm).

Everywhere in what follows we denote by X the space of entire functions $\mathcal{A}$ and by Y the space of distributions $\mathcal{A}'$, $X_R = \mathcal{A}(U_R, K)$, $Y_R = \mathcal{A}'(U_R, K)$. For the topological linear space E over the field K we use the symbol $\mathcal{L}_n(E^n, K)$ to denote the space of n_0-linear (over the field K) continuous forms $b : \ E^n \to K$.

Proposition 1.2. 1. *Let $b \in \mathcal{L}_n(X^n, K)$. Then there is $R \in \Gamma$ such that*

$$||b||_R = \sup_{\alpha} |b_{\alpha_1 \ldots \alpha_n}|_K R^{-|\alpha|} < \infty, \tag{1.1}$$

where $b_\alpha = b(e_{\alpha_1}, \ldots, e_{\alpha_n})$.

2. *For any sequence of numbers b_α satisfying condition (1.1) for a certain $R \in \Gamma$, the form*

$$b = \sum_\alpha b_\alpha e^1_{\alpha_1} \otimes \ldots \otimes e^1_{\alpha_n}/\alpha! \tag{1.2}$$

belongs to the space $\mathcal{L}_n(X^n, K)$.

Proof. 1. Let $b \in \mathcal{L}_n$. Then

$$M_R(b) = \sup_{||f^j||_R \leq 1} |b(f^1, \ldots, f^n)|_K < \infty$$

for a certain $R \in \Gamma$. Furthermore,

$$||b||_R = \sup_\alpha |b(e_{\alpha_1} a_R^{-|\alpha_1|}, \ldots, e_{\alpha_n} a_R^{-|\alpha_n|})|_K,$$

with $||e_{\alpha_j} a_R||_R = 1$ for $j = 1, \ldots, n$. Thus $||b||_R \leq M_R(b)$.

2. Consequently, for the form b defined by relations (1.2) we have

$$M_R(b) = \sup_{||f^j||_R \leq 1} |b_\alpha|_K |f^1_{\alpha_1} \cdots f^n_{\alpha_n}|_K =$$

$$= \sup_{||f^j||_R \leq 1} \sup_\alpha |b_\alpha|_K R^{-|\alpha|} |f^1_{\alpha_1}|_K R^{|\alpha_1|} \ldots |f^n_{\alpha_n}|_K R^{|\alpha_n|} \leq ||b||_R.$$

To complete the proof, we have to verify the fact that series (1.2) strongly converges in the space $\mathcal{L}_n(X^n, K)$.

When proving Proposition 1.2, we showed that

$$||b||_R = M_R(b).$$

Let us introduce an inductive topology in the space of continuous multilinear forms $\mathcal{L}_n(X^n, K)$. We set

$$\mathcal{L}_n R(X^n, K) = \{b \in \mathcal{L}_n(X^n, K) : \ ||b||_R < \infty\} = \mathcal{L}_n(X^n_R, K).$$

Proposition 1.3 *The space $\mathcal{L}_{n,R}(X^n, K)$ is a non-Archimedean Banach space.*

We set $\mathcal{L}_n(X^n, K) = \lim_{R \to \infty} \operatorname{ind} \mathcal{L}_{n,R}(X^n, K)$. By virtue of Proposition 1.3, the space $\mathcal{L}_n(X^n, K)$ is complete.

We set $|b|_R = \sup_{||f||\le 1} |b(f,\dots,f)|_K$. Then $|b|_R \le ||b||_R \le b^n|b|_R$. Consequently, the topologies defined by the norms $|\cdot|_R$ and $||\cdot||_R$ are equivalent.

By analogy with Proposition 1.2 we can prove the following proposition.

Proposition 1.4. 1. *Let $b \in \mathcal{L}_n(Y^n, K)$. Then, for all $R \in \Gamma$ we have*

$$||b||_R = \sup_\alpha |b_\alpha|_K R^{|\alpha|}/|\alpha!|_K < \infty, \tag{1.3}$$

where $b_\alpha = b(e'_{\alpha_1},\dots,e'_{\alpha_n})$.

2. *For any sequence of numbers b_α satisfying condition (1.3) for all $R \in \Gamma$, the form*

$$b = \sum_\alpha b_\alpha e_{\alpha_1} \otimes \dots \otimes e_{\alpha_n}/\alpha! \tag{1.4}$$

belongs to the space $\mathcal{L}_n(Y^n, K)$.

In the space of continuous multilinear forms $\mathcal{L}_n(Y^n, K)$ we introduce a projective topology

$$\mathcal{L}_n(Y^n, K) = \lim_{R\to\infty} \operatorname{proj} \mathcal{L}_n(Y^n_R, K).$$

Proposition 1.5. *The space $\mathcal{L}_n(Y^n, K)$ is a non-Archimedean Fréchet space.*

Proposition 1.6. $\mathcal{L}_n(Y^n, K) = \mathcal{L}'_n(X^n, K)$.

Proof. 1. Let $\tau \in \mathcal{L}_n(X^n_R, K)$, $b \in \mathcal{L}_n(Y^n, K)$. Using (1.2) and (1.4), we obtain

$$(b,\tau) = \sum_\alpha b_\alpha \tau_\alpha/\alpha!$$

Consequently,

$$|(b,\tau)|_K \le \sup_\alpha |b_\alpha|_K|\tau_\alpha|_K/|\alpha!|_K \le ||b||_R||\tau||_R.$$

2. Let us prove that series (1.2) converges in the space $\mathcal{L}_n(X^n, K)$. Let $\tau \in \mathcal{L}_{n,R}$, $R_1 \in \Gamma$, $R_1 > R$. Then

$$\left\|\sum_{|\alpha|>N} \tau_\alpha e^1_{\alpha_1} \otimes \dots \otimes e^1_{\alpha_n}/\alpha!\right\|_{R_1} \le ||\tau||_R(R/R_1)^N \to 0.$$

3. Let $b \in \mathcal{L}'_n(X^n, K)$. Then, by virtue of item 2,

$$(b, \tau) = \sum_\alpha b_\alpha \tau_\alpha / \alpha!,$$

where $b_\alpha = b(e'_{\alpha_1} \otimes \ldots \otimes e'_{\alpha_n})$. We set

$$u_\alpha = a_R^{|\alpha|}(e^1_{\alpha_1} \otimes \ldots \otimes e^1_{\alpha_n})/\alpha!.$$

Then $||u_\alpha||_R = 1$, and, consequently,

$$\sup_\alpha \frac{|b_\alpha|_K R^{|\alpha|}}{|\alpha!|_K} = \sup_\alpha |(b, u_\alpha)|_K \leq ||b||_R = \sup_{||\tau||_R \leq 1} |(b, \tau)|_K < \infty.$$

Proposition 1.7. $\mathcal{L}_n(X^n, K) = \mathcal{L}'_n(Y^n, K)$.

Proof. 1. Let us prove that series (1.4) converges in the space $\mathcal{L}_n(Y^n, K)$. Suppose that $b \in \mathcal{L}_n(Y^n, K)$, $R \in \Gamma$. Then

$$||\sum b_\alpha e_{\alpha_1} \otimes \ldots \otimes e_{\alpha_n}/\alpha!||_R \leq ||b||_{RR_1}/R_1^N \to 0, \qquad R_1 > 1.$$

2. Let $\tau \in \mathcal{L}'_n(Y^n, K)$. Then, by virtue of item 1,

$$(b, \tau) = \sum_\alpha b_\alpha \tau_\alpha / \alpha!$$

for all $b \in \mathcal{L}_n(Y^n, K)$. Since the functional τ is continuous, there exists $R \in \Gamma : \sup_{||b|| \leq 1} |(b, \tau)|_K < \infty$. We set

$$v_\alpha = a_R^{-|\alpha|}(e_{\alpha_1} \otimes \ldots \otimes e_{\alpha_n}).$$

Then $||v_\alpha||_R = 1$ and, consequently,

$$\sup_\alpha |\tau_\alpha|_K R^{-|\alpha|} = \sup_\alpha |(v_\alpha, \tau)|_K < \infty.$$

Corollary 1.1. *The spaces of the multilinear forms $\mathcal{L}_n(X^n, K)$, $\mathcal{L}_n(Y^n, K)$ are reflexive.*

2. Generalized Functions on Infinite-Dimensional Spaces

The symbol $U_{R,\rho}$ is used to denote a closed ball of radius $\rho \in \Gamma$ with respect to the norm $||\cdot||_R$, $R \in \Gamma$, in the space X : $U_{R,\rho} = \{f : ||f||_R \leq \rho\}$. We denote by $\mathcal{A}(U_{R,\rho})$ the space of analytic functions $F : U_{R,\rho} \to K$:

$$F(f) = \sum_{n=0}^{\infty} b_n(f, \ldots, f), \tag{2.1}$$

where $b_n \in \mathcal{L}_n(X^n, K)$ are symmetric. Series (2.1) converges uniformly on the ball $U_{R,\rho}$ (this is equivalent to the fact that

$$\sup_{f \in U_{R,\rho}} |b_n(f, \ldots, f)|_K \to 0, \qquad n \to \infty).$$

Note that

$$\sup_{f \in U_{R,\rho}} |b_n(f, \ldots, f)|_K = \rho^n |b_n|_R.$$

In the space $\mathcal{A}(U_{R,\rho})$ we introduce norms

$$|F|_{R,\rho} = \sup_n \sup_{f \in U_{R,\rho}} |b_n(f, \ldots, f)|_K = \sup_n \rho^n |b_n|_R,$$

$$||F||_{R,\rho} = \sup_n \rho^n ||b_n||_R.$$

Using the results of 1, we obtain

$$||F||_{R,\rho/b} \leq |F|_{R,\rho} \leq ||F||_{R,\rho}.$$

We set $\mathcal{A}_\rho(X) = \lim_{R \to \infty} \operatorname{ind} \mathcal{A}(U_{R,\rho})$. The space $\mathcal{A}_\rho(X)$ consists of functions analytic on balls of a fixed radius ρ with respect to all norms defining the topology in X. Note that

$$U_{R_2,\rho} \subset U_{R_1,\rho}, \qquad R_2 > R_1; \qquad \mathcal{A}(U_{R_2,\rho}) \supset \mathcal{A}(U_{R_1,\rho}).$$

$\mathcal{A}(U_{R,\rho})$ are non-Archimedean Banach spaces and $\mathcal{A}_\rho(X)$ are complete non-Archimedean topological linear spaces.

We denote by $\mathcal{A}_0(X)$ the space of functions analytic at zero, i.e., $\mathcal{A}_0 = \lim_{\rho\to 0} \operatorname{ind} \mathcal{A}_\rho$. Note that $\mathcal{A}_{\rho_2} \subset \mathcal{A}_{\rho_1}$, $\rho_1 < \rho_2$. The functional space $\mathcal{A}_0(X)$ is an infinite-dimensional analog of the space $\mathcal{A}_0(K^n)$.

Theorem 2.1 (on the approximation by cylindrical polynomials). *The set of polynomials dependent on a finite number of variables is dense everywhere in the space $\mathcal{A}_0(X)$.*

Proof. We shall first prove that series (2.1) converges in the space $\mathcal{A}_0(X)$. It is sufficient to prove that for ρ, $R \in \Gamma$, this series converges in the space $\mathcal{A}(U_{R,\rho})$. Indeed,

$$\left| \sum_{n=N}^{\infty} b_n \right|_{R,\rho} \le \max_{n\ge N} \sup_{f\in U_{R,\rho}} |b_n(f,\ldots,f)|_K \to 0.$$

It remains to use the fact that series (1.2) converge in the spaces $\mathcal{L}_n(X^n, K)$ for the forms b_n from (2.1).

We introduce spaces

$$\mathcal{B}_{R,\rho} = \{F \in \mathcal{A}_0 : \ ||F||_{R,\rho} < \infty\}; \qquad \mathcal{B}_\rho = \lim_{R\to\infty} \operatorname{ind} \mathcal{B}_{R,\rho}.$$

By virtue of inequalities (2.2) we have

$$\mathcal{A}_0 = \lim_{\rho\to 0} \operatorname{ind} \lim_{R\to\infty} \operatorname{ind} \mathcal{A}(U_{R,\rho}) = \lim_{\rho\to 0} \operatorname{ind} \lim_{R\to\infty} \operatorname{ind} \mathcal{B}_{R,\rho}.$$

We introduce a space of sequences

$$\Pi(K^\infty) = \{\pi = \{\pi_n\}_{n=0}^{\infty} : \ \pi_n = \{\pi_\alpha\}, \quad \alpha = (\alpha_1,\ldots,\alpha_n), \quad \pi_\alpha \in K,$$

$$\forall R,\ \rho \in \Gamma, \qquad ||\pi||_{R,\rho} = \sup_n \rho^{-n} ||\pi_n||_R < \infty, \qquad ||\pi_n||_R =$$

$$\sup_\alpha |\pi_n|_K R^{|\alpha|}\}.$$

Theorem 2.2. *The space of distributions $\mathcal{A}_0'(X)$ is isomorphic to the space $\prod(K^\infty)$.*

Proof. 1. Let $J \in \mathcal{A}_0'$. Then $\forall R$, ρ we have

$$||J||_{R,\rho} = \sup_{||F||_{R,\rho}\le 1} |J(F)|_K < \infty.$$

Let us consider functions $F_{\alpha R\rho}(f) = f_{\alpha_1} \dots f_{\alpha_n} a_R^{|\alpha|} a_\rho^{-n}$. For these functions we have

$$||F_{\alpha R\rho}||_{R,\rho} = 1, \qquad |J(F_{\alpha R\rho})|_K = |J(f_\alpha)|_K R^{|\alpha|} \rho^{-n}.$$

For the sequence $\pi = \{J(f_\alpha)\}$ we obtain $||\pi||_{R,\rho} \le ||J||_{R,\rho} < \infty$.

2. Let $\pi \in \prod(K^\infty)$. We set

$$J_\pi(F) = \sum_{n=0}^{\infty} \sum_{\alpha} \pi_\alpha b_n(e_{\alpha_1}, \dots, e_{\alpha_n}),$$

and then

$$||J_\pi||_{R,\rho} = \sup_{||F||_{R,\rho} \le 1} |J(F)|_K \le ||\pi||_{R,\rho} ||F||_{R,\rho}.$$

Recall that $Y = \lim_{R\to 0} \operatorname{ind} Y_R$, $Y_R = \mathcal{A}'(U_R)$.

The function $F: \ Y_R \to K$, is entire if series (2.1) uniformly converges on a ball of any radius $\rho \in \Gamma$ in the space Y_R. We denote by $\mathcal{A}(Y_R)$ the space of entire functions Y_R. The function $F: Y \to K$ is entire if the restriction of F to the space Y_R is an entire function for any $R \in \Gamma$. A space of entire functions is denoted by $\mathcal{A}(Y)$. This space is an infinite-dimensional analog of the space $\mathcal{A}(K^n)$. The topology in the space $\mathcal{A}(Y)$ of entire functions is defined by the system of norms

$$|F|_{R,\rho} = \sup_n \sup_{||f||_R \le \rho} |b_n(f, \dots, f)|_K.$$

The topology in $\mathcal{A}(Y)$ is similar to the topology defined by the system of norms $\{||F||_{R,\rho}) = \sup \rho^n ||b_n||_R$ (by virtue of inequalities (2.2)). The space $\mathcal{A}(Y)$ is a non-Archimedean Fréchet space.

We introduce a space of sequences

$$\Pi'(K^\infty) = \{\pi = \{\pi_n\}_{n=0}^\infty : \ \pi_n = \{\pi_\alpha\}, \qquad \pi_\alpha \in K,$$

$$\exists R, \ \rho \in \Gamma: \ ||\pi||_{R,\rho} = \sup \rho^{-n} ||\pi_n||_R < \infty, \qquad ||\pi_n||_R = \tag{2.2}$$

$$\sup |\pi_\alpha|_K R^{-|\alpha|} |\alpha!|_K\}.$$

Theorem 2.3. *The space $\mathcal{A}'(Y)$ of distributions is isomorphic to the space $\prod'(K^\infty)$.*

Proof. 1. Let $J \in \mathcal{A}'$. Then $\exists R,\ \rho \in \Gamma:\ ||J||_{R,\rho} < \infty$. Let us consider a function $F_{\alpha R\rho}(f) = a_\rho^{-n} a_R^{-|\alpha|} \alpha! f_{\alpha_1} \dots f_{\alpha_n}$. For these functions we have $||F_{\alpha R\rho}||_{R,\rho} = 1$. For the sequence $\pi = \{J(f_\alpha)\}$ we obtain

$$||\pi||_{R,\rho} = \sup_n \sup_\alpha |J(F_{\alpha R\rho})|_K \leq ||J||_{R,\rho} < \infty.$$

2. Let $\pi \in \prod'(K^\infty)$. We set

$$J_\pi(F) = \sum_{n=0}^{\infty} \sum_\alpha \pi_\alpha b(e'_{\alpha_1}, \dots, e'_{\alpha_n}),$$

and then

$$||J_\pi||_{R,\rho} = \sup_{||F||_{R,\rho} \leq 1} |J(F)|_K \leq ||\pi||_{R,\rho}.$$

We introduce spaces of infinite-dimensional differential infinite-order operators

$$\mathcal{D}_{R,\rho}(X) = \{\pi = \sum_{n=0}^{\infty} \sum_\alpha \frac{\pi_\alpha}{\alpha!} \delta^{(\alpha)}(f) :\ \pi = \{\pi_\alpha\},\quad ||\pi||_{R,\rho} < \infty\};$$

$$\mathcal{D}_0(X) = \bigcap_{\rho \in \Gamma} \bigcap_{R \in \Gamma} \mathcal{D}_{R,\rho}(X).$$

Corollary 2.1. $\mathcal{A}'_0(X) = \mathcal{D}_0(X)$.

We introduce a projective topology

$$\mathcal{A}'_0(X) = \lim_{\rho \to 0} \operatorname{proj} \lim_{R \to \infty} \operatorname{proj} \mathcal{D}_{R,\rho}(X)$$

in the space $\mathcal{A}'_0$ of distributions.

Proposition 2.1. *The space $(\mathcal{A}'_0)$ $\mathcal{A}_0$ of (generalized) test functions is reflexive.*

We define the space of differential operators $\mathcal{D}_{R,\rho}(Y)$ by analogy with $\mathcal{D}_{R,\rho}(X)$. We set $\mathcal{D}(Y) = \cup \mathcal{D}_{R,\rho}(Y)$.

Corollary 2.2. $\mathcal{A}'(Y) = \mathcal{D}(Y)$.

In the space of distributions we introduce a topology of inductive limit

$$\mathcal{A}'(Y) = \lim_{\rho \to 0} \text{ind} \lim_{R \to \infty} \text{ind}\, \mathcal{D}_{R,\rho}(Y).$$

Proposition 2.2. *The space of (generalized) test functions* $(\mathcal{A}')$ $\mathcal{A}$ *is reflexive.*

The product $J_1 \otimes J_2$ of the distributions J_1, J_2 is defined by the relation

$$(J_1 \otimes J_2,\ \varphi(f^1, f^2)) = (J_1, (J_2,\ \varphi(f^1, f^2))).$$

The convolution $J_1 \otimes J_2$ of the distributions J_1, J_2 is defined by the relation

$$(J_1 * J_2,\ \varphi(f)) = (J_1 \otimes J_2,\ \varphi(f^1 + f^2)).$$

Theorem 2.4. *The spaces of distributions* $\mathcal{A}'$ *and* $\mathcal{A}'_0$ *are topological convolution algebras.*

Proposition 2.3. *The space* $\mathcal{B}_R$ *is a non-Archimedean Banach algebra.*

Proof. Indeed,

$$||\varphi_1\varphi_2||_{R,\rho} \le \sup_n \rho^n \left|\left| \sum_{m_1+m_2=n} b_{m_1} a_{m_2} \right|\right|_R \le$$

$$\le \sup_n \rho^n \sup_{m_1+m_2=n} ||b_{m_1}||_R ||a_{m_2}||_R \le ||\varphi_1||_{R,\rho} ||\varphi_2||_{R,\rho}.$$

Corollary 2.3. *The space* $\mathcal{A}_0(X)$ *of test functions is a topological algebra.*

By virtue of Corollary 2.3, the operation of multiplication of the distribution $J \in \mathcal{A}'_0$ by the test function $\varphi \in \mathcal{A}_0$ is defined.

Proposition 2.4. *The differential operators* $\frac{\partial}{\partial f_\alpha} : \mathcal{B}_{R,\rho} \to \mathcal{B}_{R,\rho}$ *are continuous, and*

$$\left|\left| \frac{\partial \varphi}{\partial f_\alpha} \right|\right|_{R,\rho} \le \frac{R^{|\alpha|}}{\rho} ||\varphi||_{R,\rho}. \tag{2.3}$$

Proof. Indeed,

$$\frac{\partial \varphi}{\partial f_\alpha}(f) = \sum_{n=1}^{\infty} n b_n(e_\alpha, f, \ldots, f).$$

$$\left\| \sum n b_n(e_\alpha, f, \ldots, f) \right\|_{R,\rho} =$$

$$R^{|\alpha|} \sup_n \rho^{n-1} |n|_K \sup_{\|f_j\|_R \le 1} |b_n(e_\alpha a_R^{-|\alpha|}, f_1, \ldots, f_n)|_K.$$

Corollary 2.4. *The differential operators* $\frac{\partial}{\partial f} : \mathcal{A} \to \mathcal{A}_0$ *are continuous.*

By virtue of Corollary 2.4, the operation of differentiation in the space of distributions $\mathcal{A}_0'(X)$ is defined. For the multilinear form $a \in \mathcal{L}_k(Y^n, K)$ the (symmetric) differential operator

$$a(D) = \sum_\alpha a_\alpha \frac{\partial^k}{\partial f_{\alpha_1} \ldots \partial f_{\alpha_k}}.$$

Proposition 2.5. *Let* $a \in \mathcal{L}_k(Y^n, K)$. *The differential operator* $a(D) : \mathcal{B}_{R,\rho} \to \mathcal{B}_{R,\rho}$ *is continuous, and*

$$\|a(D)F\|_{R,\rho} \le \|a\|_R \rho^{-k} \|F\|_{R,\rho}. \tag{2.4}$$

Proof. By virtue of inequality (2.3),

$$\|a(D)\varphi\|_{R,\rho} \le \rho^{-k} \sup_\alpha |a_\alpha|_K R^{|\alpha|} \|\varphi\|_{R,\rho}.$$

Proposition 2.6. *Suppose that the coefficients of the differential operator*

$$b(f, D) = \sum_{k=0}^{\infty} b_k(f) a_k(D)$$

belong to the class $\mathcal{B}_{R,\rho}$, *and*

$$\lim_{k \to \infty} \|b_k\|_{R,\rho} \|a_k\|_R \rho^{-k} = 0.$$

Then the operator $b(f, D) : \mathcal{B}_{R,\rho} \to \mathcal{B}_{R,\rho}$ is continuous and we have an inequality

$$\|b(f, D)\varphi\|_{R,\rho} \le \|\varphi\|_{R,\rho} \max_k \|b_k\|_{R,\rho} \|a_k\|_R \rho^{-k}. \tag{2.5}$$

Proposition 2.7. *The space $\mathcal{A}(Y)$ is a non-Archimedean Fréchet algebra.*

Proposition 2.8. *The differentiation operators $\frac{\partial}{\partial g}$: $\mathcal{A}(Y) \to \mathcal{A}(Y)$ are continuous and we have an inequality*

$$||\frac{\partial \varphi}{\partial g_\alpha}||_{R,\rho} \leq |\alpha!|_K\, R^{-|\alpha|}\rho^{-1}||\varphi||_{R,\rho}. \tag{2.6}$$

For the multilinear (symmetric) form $a \in \mathcal{L}_k(X^k, K)$ we introduce a differential operator

$$a(D) = \sum_\alpha a_\alpha \frac{\partial^k}{\partial g_{\alpha_1} \ldots \partial g_{\alpha_k}}.$$

Proposition 2.9. *Let $a \in \mathcal{L}_k(X^k, K)$. The differential operator $a(D) : \mathcal{A}(Y) \to \mathcal{A}(Y)$ is continuous and we have inequality (2.4).*

By virtue of propositions 2.7 and 2.8, the operations of differentiation and multiplication by the test function are defined in the space of distributions $\mathcal{A}'(Y)$. By analogy we introduce spaces $\mathcal{A}_0(X, Z)$, $\mathcal{A}_0'(Y, Z)$, $\mathcal{A}(Y, Z)$, $\mathcal{A}'(Y, Z)$.

3. Laplace Transformation on Infinite-Dimensional Spaces

Proposition 3.1. *Let $g \in Y_R$, $||g||_R < (\rho b)^{-1}$. Then*

$$F_g = \exp\{(g, \cdot)\} \in \mathcal{A}(U_{R,\rho}).$$

Proof. Indeed,

$$\sup_{f \in U_{R,\rho}} |(g, f)^n|_K / |n!|_K \leq (b||g||_R\rho)^n.$$

Definition 3.1. *A (two-sided) Laplace transformation of the distribution $J \in \mathcal{A}_0'(X)$ is a function $L(J)$ on the space Y : $L(J)(g) = (J, F_g)$.*

Theorem 3.1. *The Laplace transformation* $L : \mathcal{A}_0'(X) \to \mathcal{A}(Y)$ *is an isomorphism.*

Proof. 1. Let $J \in \mathcal{A}_0'$, $L(J) = 0$. Then $(J, P_c) = 0$ for any cylindrical polynomial P_c. But by virtue of Theorem 2.1, the set of cylindrical polynomials is everywhere dense in the space $\mathcal{A}_0$. Consequently, $(J, F) = 0$ for all $F \in \mathcal{A}_0$. We have thus shown that $\operatorname{Ker} L = \{0\}$.

2. Let $J \in \mathcal{A}_0'$. We shall prove that $L(J) \in \mathcal{A}$. Indeed,

$$\sup_n |n!|_K^{-1} \sup_{||g||_R \le \rho} |(J, (g, \cdot)^n)|_K \le ||J||_{R,\rho_1} \sup_n (b\rho\rho_1)^n.$$

Choosing $\rho_1 \in \Gamma$, $\rho_1 < (b\rho)^{-1}$, we find that $||L(J)||_{R,\rho} < \infty$.

3. Let $F \in \mathcal{A}$, $F(g) = \sum b_n(g)$. We set

$$J(f_\alpha) = \frac{n!}{\alpha!} b_n(e'_{\alpha_1}, \ldots, e'_{\alpha_n}).$$

We shall show that the functional J, defined by the sequence $\{J(f_\alpha)\}$, belongs to the space $\mathcal{A}_0'$:

$$\sup_n \rho^{-n} \sup_\alpha |J(f_\alpha)|_K \, R^{|\alpha|} \le ||F||_{1/\rho,R} < \infty.$$

It remains to point out that $L(J) = F$.

By virtue of Theorem 3.1, we have the following infinite-dimensional Laplace calculus:

$$\mathcal{A}_0'(X) \xrightarrow{L} \mathcal{A}(Y), \qquad \mathcal{A}_0(X) \xleftarrow{L'} \mathcal{A}'(Y). \tag{3.1}$$

Considering functions with values in the square extension Z of the field K instead of functions with values in the field K, i.e.,

$$\mathcal{A}_0\left(\mathcal{A}\left(K^n, Z\right), Z\right), \qquad \mathcal{A}\left(\mathcal{A}'(K^n, Z), Z\right),$$

we can introduce the Fourier transform of the distribution $J \in \mathcal{A}_0'$

$$\mathcal{F}(J)(g) = (J, \exp\{\sqrt{\tau}(g, \cdot)\}).$$

Theorem 3.2. *The Fourier transformation* $F : \mathcal{A}_0' \to \mathcal{A}$ *is an isomorphism.*

By virtue of Theorem 3.2, we have the Fourier calculus

$$\mathcal{A}'_0(X,Z) \xrightarrow{\mathcal{F}} \mathcal{A}(Y,Z), \qquad \mathcal{A}_0(X,Z) \xleftarrow{\mathcal{F}'} \mathcal{A}'(Y,Z). \tag{3.2}$$

The following theorem is valid for the Fourier calculus.

Theorem 3.3. (on the properties of the Fourier transform). *Let* $J,\ J_1,\ J_2 \in \mathcal{A}'_0(X,Z)$, $\mu,\ \mu_1,\ \mu_2 \in \mathcal{A}'(Y,Z)$. *Then*

1. $\mathcal{F}(J_1 * J_2) = \mathcal{F}(J_1)\mathcal{F}(J_2)$; $\quad \mathcal{F}'(\mu_1 * \mu_2) = \mathcal{F}'(\mu_1)\mathcal{F}'(\mu_2)$.
2. $\frac{\partial}{\partial g_\alpha}\mathcal{F}(J) = \alpha!\sqrt{\tau}\mathcal{F}(f_\alpha J)$: $\quad \frac{\partial}{\partial f_\alpha}\mathcal{F}'(\mu) = \alpha!\sqrt{\tau}\mathcal{F}'(g_\alpha, \mu)$.
3. $\mathcal{F}\left(\frac{\partial J}{\partial f_\alpha}\right) = -\alpha!\sqrt{\tau} g_\alpha \mathcal{F}(J)$; $\quad \mathcal{F}'\left(\frac{\partial \mu}{\partial g_\alpha}\right) = -\alpha!\sqrt{\tau} f_\alpha \mathcal{F}'(\mu)$.

Definition 3.2. *A Gaussian distribution on a non-Archimedean space Y is a distribution of the infinite-dimensional argument $\gamma_{a,b} \in \mathcal{A}'(Y)$ which has a Laplace transformation*

$$L'(\gamma_{a,b})(x) = \exp\left\{\frac{1}{2}b(x,x) + (a,x)\right\},$$

where $a(\in Y)$ is the mean value, b (a symmetric form from $\mathcal{L}_2(X^2, K)$) is a correlation functional.

Definition 3.3. *A Feynman distribution on the non-Archimedean space Y is a generalized function of the infinite-dimensional argument $\tilde{\gamma}_{a,\sqrt{\tau}b}$ which has a Fourier transformation*

$$\mathcal{F}'(\gamma_a, \sqrt{\tau}b)(x) = \exp\left\{\tau\frac{\sqrt{\tau}}{2}b(x,x) + \sqrt{\tau}(a,x)\right\},$$

where $a(\in Y)$ is the mean value, b (a symmetric form from $\mathcal{L}_2(X^2, K)$) is a correlation functional.

Then we use integral notations

$$(\gamma_{a,b}, \varphi) = \int_Y \varphi(g)\gamma_{a,b}(dg), \qquad (\gamma_{a,\sqrt{\tau}b}, \varphi) = \int_Y \varphi(g)\gamma_{a,\sqrt{\tau}b}(dg).$$

Remark 3.1. *We use Laplace transformation rather than Fourier's for the introduction of the Gaussian distribution in order not to consider the quadratic extension which may lead to non-uniqueness in the field theories. In the non-Archimedean theory the Euclidean* *formalism described by the Gaussian distribution and by the heat conduction equation is uniquely defined. The non-uniqueness appears in the Minkowski formalism described by Feynman distribution and the Schrödinger equation.*

We could have required positive definiteness of the correlation functional $b(f,f) \in K{\cdot}K$ in definition 3.2, and (or, by immediately arriving at a non-uniqueness, $\mathrm{sign}_\tau b(f,f) = 1$). However, the Gaussian distributions of the non-Archimedean Euclidean field theory do not possess this property.

4. Linear Partial Differential Equations on Infinite-Dimensional Spaces

For the differential operator

$$b(f,D) = \sum_{k=0}^{\infty} b_k(f) a_k(D), \; b_k \in \mathcal{A}_0(X,K), a_k \in \mathcal{L}_k(Y,K),$$

let

$$\sigma_{R,\rho}(b) = \sup_k ||b_k||_{R,\rho} ||a_k||_R \rho^{-k}. \tag{4.1}$$

For the matrix differential operator $B(f,D) = (B_{ij}(f,D))$, let

$$\sigma_{R,\rho}(B) = \max_{i,j} \sigma_{R,\rho}(B_{i,j}).$$

Theorem 4.1 *Suppose that the coefficients of the matrix differential operator $B(f,D)$ belong to $\mathcal{B}_{R,\rho}$ and $\mathcal{L}_k(Y,K)$, and*

$$\lim_{k\to\infty} ||b_{ijk}||_{R,\rho} ||a_{ijk}||_R \rho^{-k} = 0.$$

Then on the interval $|t|_K < (b\sigma_{R,\rho})^{-1}$, where b is the coefficient of the inequality (1.5),Ch. 1, there exists a unique solution of the Cauchy problem

$$\frac{\partial}{\partial t} F(t,f) = B(f,D) F(t,f), \; F(0,f) = \Phi(f), \tag{4.2}$$

that is contained in the space $\mathcal{B}_{R,\rho}$ *and is analytic with respect to* t.

Indeed, by (2.5),

$$\|\exp\{tB(f,D)\Phi\}\|_{R,\rho} \leq C\sup_n(|t|_K b)^n \|B^n(f,D)\Phi\|_{R,\rho} \leq$$

$$C\sup_n(|t|_K b\sigma_{R,\rho}(B))^n \|\Phi\|_{R,\rho}.$$

Corollary 4.1 *The Cauchy problem (4.2) is uniquely locally solvable in the space* $\mathcal{B}_{R,\rho}$ *for any differential operator of finite order with coefficients* $b_{ijk} \in \mathcal{B}_{R,\rho}$ *and* $a_{ijk} \in \mathcal{L}_k$.

Corollary 4.2 *Let* $b_j(f,D), j = 0, ..., m-1$, *be differential operators with coefficients* $b_{jk} \in \mathcal{B}_{R,\rho}$ *and* $a_{jk} \in \mathcal{L}_k$ *such that*

$$\lim_{k\to\infty} \|b_{jk}\|_{R,\rho}\|a_{jk}\|_R \rho^{-k} = 0.$$

Then in the space $\mathcal{B}_{R,\rho}$ *there is a unique solution of the Cauchy problem*

$$\frac{\partial^m F}{\partial t^m}(t,f) + \sum_{j=o}^{m-1} b_j(f,D)\frac{\partial^j F}{\partial t^j}(t,f) = 0, \tag{4.3}$$

$$\frac{\partial^j F}{\partial t^j}(0,f) = \Phi_j(f), j = 0, ..., m-1, \tag{4.4}$$

on the interval $|t|_K < (b\gamma)^{-1}, \gamma = \max_j \sigma_{R,\rho}(b_j)$, *that is analytic with respect to* t.

Corollary 4.3 *Let* $B(f,D)$ *be a differential operator of finite order. Then the Cauchy problem (4.2) is locally solvable in the space* $\mathcal{A}_0(X,K)$. *The solution is unique in the class of such functions.*

Example 4.1 (the heat equation). Suppose that

$$\sup_\alpha |b\alpha|_K R^{2|\alpha|}/|\alpha!|_K < \infty$$

for all $R \in \Gamma$. Then the Cauchy problem for the equation

$$\frac{\partial F}{\partial t}(t,f) = \sum_{|\alpha|=0}^{\infty} b_\alpha \frac{\partial^2 F}{\partial f_\alpha^2}(t,f) \tag{4.5}$$

is solvable on the interval

$$|t|_K < \rho^2 \Big(b \sup_{\alpha} |b_\alpha|_K R^{|\alpha|} / |\alpha!|_K \Big)^{-1}$$

for $\phi \in \mathcal{B}_{R,\rho}$.

The Cauchy problem for the heat equation with potential

$$\frac{\partial F}{\partial t}(t,f) = \sum_{|\alpha|=0}^{\infty} b_\alpha \frac{\partial^2 F}{\partial f_\alpha^2}(t,f) + V(f)F(t,f) \tag{4.6}$$

is solvable on the interval

$$|t|_K < b^{-1} \max \Big[||V||_{R,\rho}, \max_{\alpha} |b_\alpha|_K R^{|\alpha|} / |\alpha!|_K \rho^{-2} \Big].$$

It can be proved analogously that in $\mathcal{A}_0(X, Z)$ the Cauchy problem is locally solvable for the Schrödinger equation with potential

$$\frac{h}{\sqrt{\tau}} \frac{\partial \Psi}{\partial t}(t,f) = \frac{h^2}{2} \sum_{|\alpha|=0}^{\infty} b_\alpha \frac{\partial^2 \Psi}{\partial f_\alpha^2}(t,f) - V(f)\Psi(t,f). \tag{4.6}$$

Thus, in the non-Archimedean case the heat equation and the Schrödinger equation are locally solvable in $\mathcal{A}_0$ for analytic potentials with arbitrarily rapid growth.

Applying the Laplace operator $(L')^{-1}$ to the Cauchy problems (4.2) and (4.3),(4.4), we get the Cauchy problem

$$\frac{\partial}{\partial t} G(t,g) = B(g,D)G(t,g), G(0,g) = \Psi(g), \tag{4.8}$$

in the space $A'(Y, K)$ of distributions, where

$$B^{ij}(D,g) = \sum_{k=0}^{N} b_k^{ij}(D) a_k^{ij}(g), b_k^{ij} \in \mathcal{A}_0, a_k^{ij} \in \mathcal{L}_k;$$

and the Cauchy problem

$$\frac{\partial^m G}{\partial t^m}(t,f) + \sum_{j=o}^{m-1} b_j(D,g) \frac{\partial^j G}{\partial t^j}(t,g) = 0, \tag{4.9}$$

$$\frac{\partial^j G}{\partial t^j}(0,g) = \Psi_j(g), j = 0, ..., m-1, \tag{4.10}$$

where

$$b_j(D,g) = \sum_{k=0}^{N} b_{jk}(D)a_{jk}(g), b_{jk} \in \mathcal{A}_0, a_{jk} \in \mathcal{L}_k.$$

Corollary 4.4. *The Cauchy problems (4.8) and (4.9),(4.10) are locally solvable in the space $\mathcal{A}'(Y,K)$ of distributions. The solution is unique in the class of functions that assume values in $\mathcal{A}'(Y,K)$ and are analytic with respect to t.*

The next result follows from Propositions 2.7 and 2.9.

Proposition 4.1 *The differential operator of a finite order*

$$b(g,D) = \sum_{k=0}^{N} b_k(g)a_k(D), b_k \in \mathcal{A}(Y,K), a_k \in \mathcal{L}_k(X,K),$$

is continuous in the space $\mathcal{A}'(Y,K)$, and inequality (2.5) holds.

The constants $\sigma_{R,\rho}(b)$ are defined by (4.1).

Theorem 4.2 *Suppose that the coefficients of the matrix differential operator $B(f,D) = (B_{ij}(f,D))$ (of the differential operator $b_j(f,D)$) belong to $\mathcal{A}(Y,K)$ and $\mathcal{L}_k(X,K)$, and*

$$\kappa = \overline{\lim}_{R,\rho\to\infty}\sigma_{R,\rho}(B) \text{ (or } \sigma_{R,\rho}(b)) < \infty.$$

Then the Cauchy problem (4.2), ((4.3),(4.4)) is solvable in the space $\mathcal{A}(Y,K)$ on the interval $|t|_K < (b\kappa)^{-1}$. The solution, which is analytic with respect to t, is unique in this space and depends continuously on the initial conditions.

It suffices to use (2.5) for the proof.

Corollary 4.5. *Suppose that $B(D) = (B_{ij}(D))$, where $B_{ij}(D) = \sum_{k=1}^{N} a_{ijk}(D)$ with $a_{ijk} \in \mathcal{L}_k(X,K)$. Then the Cauchy problem*

$$\frac{\partial}{\partial t}G(t,g) = B(D)G(t,g), G(0,g) = \Psi(g), \tag{4.11}$$

is globally solvable in the space $\mathcal{A}(Y,K)$.

Corollary 4.6. *There exists a unique fundamental solution of the Cauchy problem (4.11) in the space $\mathcal{A}'(Y,K)$ of distributions.*

The heat equation (4.5) is solvable in $\mathcal{A}$ under essentially weaker restrictions on the coefficients of the Laplace operator :

$$\sup_{\alpha} |b|_{\alpha} R^{-|\alpha|} < \infty.$$

The fundamental solution of (4.5) is a Gaussian distribution :

$$F(t,f) = \int_Y F(0,y)\gamma_{f,2tB}(\mathrm{d}y),$$

where B is a diagonal matrix $B = (b_{\alpha})$.

The fundamental solution of the Schrödinger equation is a Feynman distribution. If $V = 0$, then

$$\Psi(t,f) = \int_Y \Psi(0,y)\gamma_{f,th\sqrt{T}B}(\mathrm{d}y).$$

Let us denote by $\mathcal{A}^0(\mathcal{U}_t, Y)$ the space of analytic functions $\omega : \mathcal{U}_t \to Y$, $\mathcal{U}_t = \{s \in K : |s|_K \le t\}$, with the condition $\omega(0) = 0$. The Feynman-Kac Formulas hold for the heat equation and the Schrödinger equation with potential:

$$F(t,f) = \int\limits_{\mathcal{A}^0(\mathcal{U}_t,Y)} F(0,\omega(t)) \exp\{\int_0^t V(\omega(s))\mathrm{d}s\} W_{f,B}(\mathrm{d}\omega(s)),$$

$$\Psi(t,f) = \int\limits_{\mathcal{A}^0(\mathcal{U}_t,Y)} \Psi(0,\omega(t)) \exp\{-\frac{\sqrt{T}}{h}\int_0^t V(\omega(s))\mathrm{d}s\} \Phi_{f,B}(\mathrm{d}\omega(s)),$$

where $W_{f,B}$ and $\Phi_{f,B}$ are the corresponding Wiener and Feynman distributions on the space $\mathcal{A}^0(\mathcal{U}_t, Y)$. The indefinite integral of an analytic function on the field K is defined in the standard way with the aid of integrals: $\int_a^b t^n = \frac{b^{n+1}-a^{n+1}}{n+1}$. However, the proof of these formulas is a fairly complicated problem, and it is not possible for us to investigate this problem here.

Chapter IV

Quantum Mechanics for Non-Archimedean Wave Functions

The quantum mechanics in which wave functions assume values in quadratic extensions of non-Archimedean fields was suggested in [72], [64], [58]. Representations of Bargmann–Fock and Schrödinger were constructed in these papers in the spaces of functions square summable with respect to the non-Archimedean Gauss and Lebesgue distributions. For quantization use was made of the calculus of pseudo-differential operators acting in the spaces of functions of a non-Archimedean argument with non-Archimedean values (this calculus is a natural generalization of the calculuses of pseudo-differential operators in an infinite-dimensional case [43], [45], [47], [48], [53], [54] and on a superspace [49], [50], [52], [55], [56]). What is common for all these calculuses is the absence of the Lebesgue measure. Here everything is realized in the sense of the theory of distributions. The principle of correspondence of the quantum mechanics and the classical non-Archimedean-valued mechanics was proved in the framework of calculus of pseudo-differential operators (the non-Archimedean deformation parameter was considered).

Papers [64], [57] studied linear evolutional equations with variable coefficients in the spaces of analytic functions. Here the existence and uniqueness theorems were obtained as well as those of the continuous dependence on the initial conditions. These results were used to investigate equations of the non-Archimedean mathematical physics and sufficient conditions for solvability of the equations of Schrödinger, Heisenberg, Liouville and of heat condition were obtained [64].

Paper [64] suggested a model of statistical quantum mechanics with non-Archimedean distribution functions and studied the corresponding evolution equations.

In connection with the development of quantum mechanics in which wave functions assume values in non-Archimedean fields, a number of new mathematical problems were posed such as the theory of non-Archimedean Hilbert spaces [64], [72] the spectral theory of operators in these spaces [64] (the advances here are minimal), the probability theory in which probabilities can belong not only to the field of real numbers but also to the fields of p-adic numbers [67]. The latter theory seems to be the most interesting for applications (see Ch. 6).

Originally the non-Archimedean-valued quantum mechanics appeared as a mathematical formalism generalizing the formalism of the ordinary quantum mechanics. Later on investigations were carried out [76], [21] in whose framework physical realizations of the proposed mathematical formalism were suggested. First, it is a statistical interpretation in the framework of the p-adic probability theory. From this point of view the p-adic-valued quantum models are standard quantum models in which additional quantum states are considered which have a p-adic probability of realization. Second, it is [65] an attempt to interpret the p-adic values observed as infinite quantities, in cosmological models first of all. With this approach, it is possible to consider systems with an infinite number of particles in a fixed state, infinite space-time intervals.

It stands to reason that we distinguish between two problems, namely, the problem of constructing mathematical formalism for quantum theories with non-Archimedean-valued functions and the problem of physical interpretation of this formalism. The suggested interpretation is, evidently, not unique and other applications of the constructed mathematical formalism are possible.

1. Schrödinger and Bargmann–Fock Representations in Non-Archimedean Quantum Mechanics

The states of the non-Archimedean quantum mechanics system in Schrödinger's representation are described by functions from the space $L_2(K^n, \mathrm{d}x)$. The observers are associated with the operators in $L_2(K^n, \mathrm{d}x)$ symmetric with respect to (5.1),Chapter 2, the mean value of the observers $\hat{a}$ in the state φ is defined by the relation $\bar{a}_\varphi = (\hat{a}, \varphi, \varphi)$ and the dynamics is described by Schrödinger's equation $(-h/\sqrt{\tau})\dot{\psi}(t) = \hat{H}\psi(t)$, where $\hat{H}$ is the Hamiltonian.

Remark 1.1 (on the positive definiteness of the quantum mechanics operators). *It is evidently impossible to introduce the concept of positiveness for* $\hat{H}$. *We could have required that* $\operatorname{sign}_\tau \bar{H}_\varphi \geq 0$, *but this requirement is not satisfied for the simplest non-Archimedean models. The Hamiltonian of the harmonic oscillator* $\hat{H} = \frac{1}{2}\left(-\frac{\partial^2}{\partial x^2} + x^2 - 1\right)$ *has eigenvalues* $\lambda_n = n$ *(the eigenfunctions are Hermitian functions* φ_n*),* $\bar{H}_{\varphi_n} = n$. *Let* $K = Q_2$, $\tau = 3$, $n = 3$. *Then Hilbert's symbol [17]* $(3,3) = -1$ *and, consequently,* $\operatorname{sign}_3 3 = -1$. *On the other hand, if the field* K *contains* $i = \sqrt{-1}$, *then* $\operatorname{sign}_\tau(-1) = 1$ *and, consequently, we should have considered the operator* $(-I)$ *to be positive.*

Remark 1.2 (on the self-adjointness of quantum mechanics operators). *Since* $L_2'(K^n, \mathrm{d}x) \neq L_2(K^n, \mathrm{d}x)$, *the adjoint operator* $\hat{a}^* : L_2' \to L_2'$ *and the self-adjointness of the operator cannot be defined in a standard way. We can introduce an analog of the concept of self-adjointness by means of the spectral expansion* $\hat{a} = \int_K \lambda \, \mathrm{d}E_\lambda$, *where the operator integral is understood in a weak sense* $(\hat{a}f, g) = \int \lambda \, \mathrm{d}(E_\lambda f, g)$ *and the integral with respect to the field* K *in the sense of distributions. If* $(E_\lambda f, g) \in \mathcal{A}_\rho'$, *then the functions* $u(\hat{a})$, $u \in \mathcal{A}_\rho$ *of the operator* $\hat{a}$ *are defined:* $(u(\hat{a})f, g) = \int u(\lambda) \, \mathrm{d}(E_\lambda f, g)$. *For instance,* $(E_\lambda f, g) \in \mathcal{A}_1'$ *for the Hamiltonian* $\hat{H}$ *of the harmonic oscillator. Note that the operator* $\hat{H}$ *is bounded and* $||\hat{H}|| = 1$.

Remark 1.3 (on the probability interpretation of the non-Archimedean quantum mechanics). *The generalized function* $\mathrm{P}_\varphi(\lambda) =$

$|E_\lambda \varphi|^2$ *is called a distribution function of* $\hat{a}$ *being observed in the state* φ. *In particular,* $\bar{a}_\varphi$ *can be represented as a probability mean* $\bar{a}_\varphi = \int \lambda \, \mathrm{d}\mathrm{P}_\varphi(\lambda)$. *It is impossible to calculate the probability of the value of the observed function* $\hat{a}$, *measured in the state* φ, *belonging to the Borel set* Δ *(in any event, in the framework of our theory of distributions). We can give the following definition: the values of the observed* $\hat{a}$ *measured in the state* φ *with an individual probability lie in the interval* $(\delta_1, \delta_2]$, $\delta_i \in \Gamma$, *if* $\mathrm{P}_\varphi(\lambda) \in \mathcal{A}'\delta_2 \backslash \mathcal{A}'_{\varphi_1}$. *For instance, the values of the oscillator energy lie, with the individual probability, on a unit interval.*

Remark 1.4 (prequantization). *If the group* Γ *is discrete, then, already at the classical level, the coordinates* q_j, *the momenta* p_j, *the mass* m, *and the energy* H *assume discrete values. Thus the transition from a classical real system to a classical non-Archimedean system leads to the discretization of the quantities being observed.*

If the distribution function of the observed $\hat{a}$ in the state φ is Gaussian $\mathrm{P}_\varphi(\mathrm{d}\lambda) = \gamma_{\bar{a}\varphi}, \ \mathcal{B}_\varphi$, then there is a probability that $\hat{a} \in [c, d]$, $\mathrm{P} = \int_c^d \mathrm{P}_\varphi(\mathrm{d}\lambda)$.

The PDO $\hat{a}$ with the (qp) symbol $a(q, p)$ is defined by the relation $(\varphi \in W, \ a \in \mathcal{A})$:

$$\hat{a}(\varphi)(q) = \int_{K^{2n}} a(q, p_1)\varphi(q_1) \mathrm{e}^{\frac{\sqrt{\tau}}{h}(p_1, q-q_1)} \, \mathrm{d}q_1 \mathrm{d}p_1.$$

In particular, the operators of the coordinates $\hat{q}_j(\varphi)(q) = (q)_j \varphi(q)$ and of the momentum $\hat{p}_j(\varphi)(q) = (h/\sqrt{\tau}) \frac{\partial \varphi}{\partial q_j}(q)$ are PDO with symbols q_j and p_j respectively. This follows from the formula

$$\varphi(q) = \int_{K^{2n}} \varphi(q_1) \mathrm{e}^{\frac{\sqrt{\tau}}{h}(p_1, q-q_1)} \, \mathrm{d}q_1 \mathrm{d}p_1. \tag{1.1}$$

Let us prove this formula:

$$\int J(\varphi)(q_1) \exp\Big\{ - \frac{|q_1|^2}{2} + \frac{\sqrt{\tau}}{h}(p_1, q - q_1) \Big\} \, \mathrm{d}q_1 \mathrm{d}p_1 =$$

$$= \exp\Big\{\frac{\tau}{2h^2}(B^{-1}a, a)\Big\}\times$$

$$\int J(\varphi)(q_1)\exp\Big\{-\frac{1}{2}\Big(B\Big(x-\frac{\sqrt{\tau}}{h}B^{-1}a\Big), x-\frac{\sqrt{\tau}}{h}B^{-1}a\Big)\Big\}\,\mathrm{d}x,$$

where $x = q_1 \oplus p_1$, $a = 0 \oplus q$, $B = \begin{pmatrix} 1 & \sqrt{\tau}/h \\ \sqrt{\tau}/h & 0 \end{pmatrix}$. Using the Parseval equality, we have

$$\exp\Big\{-\frac{|q|^2}{2}\Big\}\int L^{-1}(J(\varphi))\otimes\delta(\mathrm{d}x)\times\exp\Big\{\frac{1}{2}(B^{-1}x, x)\Big\}\times$$

$$\exp\Big\{\frac{\sqrt{\tau}}{h}(x, B^{-1}a)\Big\} = \exp\Big\{-\frac{|q|^2}{2}\Big\}\int L^{-1}(J(\varphi))(\mathrm{d}p)\mathrm{e}^{(p,q)}.$$

Theorem 1.1 (the composition formula). *Let $\hat{a}$, $\hat{a}_1$, $\hat{a}_2$ be PDO with symbols a, a_1, $a_2 \in \mathcal{A}(K^{2n})$, $\hat{a} = \hat{a}_1 \circ \hat{a}_2$. Then*

$$a(q,p) = a_1 * a_2(q,p) = \int_{K^n} 2a_1(q,p_1)\,a_2(q_1,p)\mathrm{e}^{-\frac{\sqrt{\tau}}{h}(p-p_1,q-q_1)}\,\mathrm{d}q_1\mathrm{d}p_1. \tag{1.2}$$

Proof.

$$\hat{a}(\varphi)(q) = \int a_1(q,p_1)\int a_2(q_1,p)\varphi(q_2)\times$$

$$\exp\Big\{\frac{\sqrt{\tau}}{h}[(p_1, q-q_1)+(p, q_1-q_2)]\Big\}\mathrm{d}q_1\mathrm{d}p_1\mathrm{d}q_2\mathrm{d}p.$$

Theorem 1.2. *Suppose that the conditions of Theorem 1.1 are satisfied. Then*

$$a(q,p) = a_1\Big(q,\, p+\frac{h}{\sqrt{\tau}}\frac{\partial}{\partial q_1}\Big)\,a_2(q_1,p)\Big|_{q_1=q}, \tag{1.3}$$

$$a(q,p) = \mathrm{e}^{\frac{h}{\sqrt{\tau}}(\frac{\partial}{\partial p_1},\frac{\partial}{\partial q_1})}a_1(q,\, p_1)\,a_2(q_1,\, p)\,|_{q_1=q,p_1=p}. \tag{1.4}$$

Proof. It is sufficient to prove formula (1.3) for the functions $a_1 = \alpha_1(q)\,\alpha_2(p)$, $a_2 = \beta_1(q)\,\beta_2(p)$. Using (1.2), we get

$$a(q,p) = \alpha_1(q)\beta_2(p)\int\alpha_2(p_1)\beta_1(q_1)\mathrm{e}^{-\frac{\sqrt{\tau}}{h}(p_1-p,q_1-q)}\,\mathrm{d}q_1\mathrm{d}p_1 =$$

$$= \alpha_1(q)\beta_2(p) \sum_{|\alpha|=0}^{\infty} \frac{\alpha_2^{(\alpha)}(p)}{\alpha!} \int \xi^\alpha \beta_1(q_1) e^{\frac{\sqrt{\tau}}{h}(\xi, q-q_1)} \mathrm{d}q_1 \mathrm{d}\xi.$$

Using (1.1), we obtain

$$a(q,p) = \sum_{|\alpha|=0}^{\infty} \frac{1}{\alpha!} \left(\frac{h}{\sqrt{\tau}}\right)^{|\alpha|} \frac{\partial^\alpha a_1}{\partial p^\alpha}(q,p) \frac{\partial^\alpha a_2}{\partial q^\alpha}(q,p) = (1.3).$$

Theorem 1.3 (the principle of correspondence of the classical and quantum mechanics). *Let* $a_1,\ a_2 \in \mathcal{A}$. *Then*

1. $\lim_{h\to 0} a_1 * a_2(q,p) = a_1(q,p)a_2(q,p)$,
2. $\lim_{h\to 0} \frac{\sqrt{\tau}}{h}[a_1, a_2]_*(q,p) = \{a_1, a_2\}(q,p)$, *where* $[a_1, a_2]_* = a_1 * a_2 - a_2 * a_1$, $\{a_1, a_2\}$ *are Poisson's brackets.*

The proof of the theorem is based on the composition formula (1.4).

Heisenberg's equation on a non-Archimedean space

$$\frac{h}{\sqrt{\tau}} \frac{\mathrm{d}\hat{a}}{\mathrm{d}t} = [\hat{a}, \hat{H}] \tag{1.5}$$

can be written as an equation for the PDO symbol:

$$\frac{\mathrm{d}a(t,q,p)}{\mathrm{d}t} = \sum_{n=1}^{\infty} \left(\frac{h}{\sqrt{\tau}}\right)^{n-1} \frac{1}{n!} \left(\frac{\partial^n a}{\partial p^n}\frac{\partial^n H}{\partial q^n} - \frac{\partial^n a}{\partial q^n}\frac{\partial^n H}{\partial p^n}\right)(t,q,p).$$

By virtue of the correspondence principle, as $h \to 0$, Heisenberg's equation turns into *Liouville's equation.*

The representation of Bargmann–Fock in the non-Archimedean quantum mechanics is realised in the space $F_2(Z^n, \gamma)$ by means of the PDO calculus with Wick symbols

$$\hat{a}(\varphi)(q) = \int_{Z^{2n}} a(z, \bar{v})\varphi(v) e^{\frac{1}{h}(\bar{v}, z-v)} \mathrm{d}v\mathrm{d}\bar{v},$$

$$a \in \mathcal{A}(Z^{2n}, Z), \qquad \varphi \in \mathcal{A}(Z^n, Z). \tag{1.6}$$

The creation operator $\hat{a}_j^* \varphi = z_j \varphi$ and the annihilation operator $a_j \varphi = h \frac{\partial \varphi}{\partial z_j}$ are PDO with Wick symbols z_j and $\bar{z}_j$. This follows from the formula

$$\varphi(z) = \int_{Z^{2n}} \varphi(v) \mathrm{e}^{\frac{1}{h}(\bar{v}, z-v)} \, dv d\bar{v}. \tag{1.7}$$

For the Gauss distributions the formula for integration by parts yields $(\hat{a}_j \varphi, \psi) = (\varphi, \hat{a}_j^* \psi)$, φ, $\psi \in \mathcal{A}$. The composition formulas for the PDO with Wick symbols have the form

$$a = a_1 * a_2(z, \bar{z}) = \int_{Z^{2n}} a_1(z, \bar{v}) a_2(v, \bar{z}) \mathrm{e}^{-\frac{1}{h}(z-v, \bar{z}-\bar{v})} \, dv d\bar{v}; \tag{1.8}$$

$$a(z, \bar{z}) = a_1 \left(z, \ \bar{z} + h \frac{\partial}{\partial v} \right) a_2(v, \bar{z}) \Big|_{v=z}; \tag{1.9}$$

$$a(z, \bar{z}) = \mathrm{e}^{h\left(\frac{\partial}{\partial v}, \frac{\partial}{\partial \bar{v}}\right)} a_1(z, \bar{v}) a_2(v, \bar{z}) \Big|_{\substack{v=z \\ \bar{v}=\bar{z}}}. \tag{1.10}$$

Theorem 1.4. *Suppose that* a, a_1, $a_2 \in \mathcal{A}(Z^{2n}, Z)$. *Then relations* (1.8)–(1.10) *are satisfied.*

Using (1.10), we obtain the following theorem.

Theorem 1.5 (the principle of correspondence of the classical and quantum mechanics). *Let* a_1, $a_2 \in \mathcal{A}$. *Then*

1. $\lim_{h \to 0} a_1 * a_2(z, \bar{z}) = a_1(z, \bar{z}) a_2(z, \bar{z})$;
2. $\lim_{h \to 0} \frac{1}{h} [a_1, a_2]_*(z, \bar{z}) = \{a_1, a_2\}(z, \bar{z})$, where

$$\{a_1, a_2\} = \sum_{j=1}^{n} \left(\frac{\partial a_1}{\partial \bar{z}_j} \frac{\partial a_2}{\partial z_j} - \frac{\partial a_1}{\partial z_j} \frac{\partial a_2}{\partial \bar{z}_j} \right).$$

Theorem 1.6 (on the equivalence of the representations of Schrödinger and Bargmann–Fock). *Let* $\sqrt{2} \in Z$. *Then the representations of Schrödinger and Bargmann–Fock are equivalent.*

If $\sqrt{2} \notin Z$ (say, $Z = \mathbb{Q}_2(\sqrt{3})$), then the question of the equivalence of representations of Schrödinger of Bargmann–Fock remain open.

2. Non-Archimedean Quantum Statistical Mechanics

In the classical statistical scheme [14] we can consider either Liouville's equation for the distribution of probabilities $\mathcal{D}_t(d\Omega)$, $\Omega = (q,p)$, in the space of distributions

$$\frac{\partial \mathcal{D}_t(d\Omega)}{\partial t} = \{\mathcal{D}_t(\mathrm{d}\Omega), H(\Omega)\}, \tag{2.1}$$

or Liouville's equation for the density of the probabilities distribution $\mathcal{D}(t,\Omega)$ in the space $L_2(K^n, \mathrm{d}x)$. Passing from Liouville's equation in $L_2(K^n, \mathrm{d}x)$ to Liouville's equation in $L_2(K^n, \nu)$: $\mathcal{D}(t,\Omega) = \mathrm{P}(t,\Omega)\mathrm{e}^{-|\Omega|^2/2}$, we obtain

$$\frac{\partial \mathrm{P}}{\partial t}(t,\Omega) = \{\mathrm{P}(t,\Omega), H(\Omega)\} + u(\Omega)\mathrm{P}(t,\Omega), \tag{2.2}$$

where

$$u(\Omega) = \sum_{j=1}^{n} \left(\frac{\partial H}{\partial p_j} q_j - \frac{\partial H}{\partial q_j} p_j \right).$$

In the quantum statistical scheme [14] Neumann's statistical operator $\mathcal{D}_t$ satisfies Heisenberg's equation (1.5) and its symbol satisfies equation (1.6).

Just as for the Hamiltonian, it is evidently impossible to introduce any definition of the possible definiteness for the statistical operator $\mathcal{D}_t$.

Suppose, for instance (see [14]), that the set of admissible states at the moment $t = 0$ is represented by the (normalized) functions $(\ldots, \psi_n(x), \ldots)$ and the probability of the nth state is equal to ω_n, i.e., $\Sigma\omega_n = 1$. Then, for the kernel $\mathcal{D}_t(x,x')$ of the statistical operator $\hat{\mathcal{D}}_t$ we have

$$\mathcal{D}_t(x,x') = \sum_n \omega_n \psi_n(t,x)\overline{\psi_n(t,x)},$$

where $\psi_n(t,x)$ are solutions of Schrödinger's equation for the initial conditions of ψ_n. Consequently,

$$\int h(x)\mathcal{D}_t(x,x')\bar{h}(x)\,\mathrm{d}x\mathrm{d}x' = \sum_n \omega_n \left| \int h(x)\psi_n(t,x)\,\mathrm{d}x \right|^2 .$$

But $\operatorname{sign}_\tau \Sigma\omega_n |\int h(x)\psi(t,x)\,\mathrm{d}x|^2$ is, in general, equal to (-1).

In the framework of the theory of Gaussian integrals, the canonical Gibbs' distribution for the Hamiltonian quadratic functions is correctly defined: $\mathcal{D}(\mathrm{d}\Omega) = \exp\{-\beta H(\Omega)\}\,\mathrm{d}\Omega$.

3. The Theorems of the Existence and Uniqueness of Solution of Linear Partial Differential Equations on Non-Archimedean Spaces

We set $\tau_\rho(a) = \sup_\beta(\rho^{-|\beta|}||a_\beta||_\rho)$ for the differential operator (of an infinite order) with variable coefficients $a(x,D) = \sum_0^\infty a_\beta(x)\frac{\partial^\beta}{\partial x^\beta}$ and $\sigma_\rho(A) = \max \sigma_\rho(a_{ij})$ for the matrix differential operator $A(x,D) = (a_{ij}(x,D))$.

Theorem 3.1. *Suppose that $a_{ij\beta}$, $\varphi \in \mathcal{A}_\rho$, with*

$$\lim_{|\beta|\to\infty} \rho^{-|\beta|}||a_\beta||_\rho = 0 \tag{3.1}$$

Then, in the space $\mathcal{A}_\rho$ on the interval $|t|_K < 1/b\sigma_\rho(A)$, there is a solution of Cauchy's problem, which is a unique solution analytic with respect to t:

$$\frac{\partial u}{\partial t}(t,x) = A(x,D)u(t,x), \qquad u(0,x) = \varphi(x). \tag{3.2}$$

The solution has the form

$$u(t,x) = \mathrm{e}^{tA(x,D)}\varphi(x). \tag{3.3}$$

Condition (3.1) is satisfied, in particular, for all finite-order differential operators.

Example 3.1 (the equation of the non-Archimedean diffusion):

$$\frac{\partial u}{\partial t}(t,x) = \sum_{i,j=1}^{n} a_{ij}(x) \frac{\partial^2 u}{\partial x_i \partial x_j}(t,x).$$

This equation can be solved on the interval $|t|_K < \rho^2/b \max_{i,j} ||a_{ij}||_\rho$ for the initial conditions of the class $\mathcal{A}_\rho$.

Corollary 3.1. *Suppose that the conditions of Theorem* 3.1 *are satisfied. Then, in the space* $\mathcal{A}'_\rho$ *on the interval* $|t|_K < 1/b\sigma_\rho(A)$ *there is a solution of Cauchy's problem which is a unique solution analytic with respect to* t*:*

$$\frac{\partial g}{\partial t}(t,x) = A^*(x,D)g(t,x), \qquad g(0,x) = \psi(x), \tag{3.4}$$

where

$$A^*(x,D) = \sum_{|\beta|=0}^{\infty} (-1)^{|\beta|} \frac{\partial^\beta}{\partial x^\beta} a_\beta(x).$$

Corollary 3.2. *Let* $A(x,D)$ *be a finite-order differential operator. Then, in the space of test functions* $\mathcal{A}_0$ *there is a unique, analytic with respect to* t*, solution of Cauchy's problem* (3.2).

Note that there is, in general, no interval of existence of a solution that would hold in the whole space of test functions.

Theorem 3.2. *Suppose that the finite-order matrix differential operator* $A(x,D)$ *satisfies the condition*

$$\kappa = \overline{\lim_{\rho\to 0}} \sigma_\rho(A) < \infty. \tag{3.5}$$

Then Cauchy's problem (3.2) *has a unique, analytic with respect to* t*, solution in the space* $\mathcal{A}$ *on the interval* $|t|_K < 1/\kappa b$.

The solution is defined by (3.3) and, consequently, the following theorem is valid.

Theorem 3.3. *Suppose that the conditions of Theorem* 3.2 *are satisfied. Then, in the space* $\mathcal{A}'$ *of distributions, there is a unique, analytic with respect to* t*, solution of Cauchy's problem* (3.4) *on the interval* $|t|_K < 1/\kappa b$.

4. Solvability of the Schrödinger, Heisenberg and Liouville Equations in Non-Archimedean Mechanics

Suppose that the function $v \in \mathcal{A}_\rho$ in the space $\mathcal{A}_\rho$ is defined by Schrödinger equation

$$-\frac{h}{\sqrt{\tau}}\frac{\partial\psi}{\partial t}(t,x) = \frac{h^2}{2\tau}\Delta\psi(t,x) + v(x)\psi(t,x). \tag{4.1}$$

It follows from Theorem 3.1 that Cauchy's problem for equation (4.1) is solvable on the interval $|t|_K < \delta(h,b,\tau,v,\rho)$ [69]. By virtue of Corollary 3.2, Cauchy's problem (4.1) is locally solvable in the space of test functions $\mathcal{A}_0$. Let us consider equation (4.1) in the space $L_2(K^n, dx)$. We pass to the equation in the space $L_2(K^n,\nu)$:

$$-\frac{h}{\sqrt{\tau}}\frac{\partial\varphi}{\partial t}(t,x) = \frac{h^2}{2\tau}\sum_{j=1}^{n}\left(\frac{\partial}{\partial x_j} - x_j\right)^2\psi(t,x) + v(x)\psi(t,x).$$

The space $L_2(K^n,\nu)$ is injected into the space $\mathcal{A}_\rho$, $\rho < \sqrt{b/|2|_K}$. By virtue of Theorem 3.1, on the subspace $\mathcal{A}_\rho$ Cauchy's problem for equation (4.1) is solvable on the interval $|t|_K < \delta(h,b,\tau,v,\rho)$ [69].

Let us consider Liouville's equation (2.1). We set

$$\mathcal{L} = \sum_{j=1}^{n}\left(\frac{\partial H}{\partial p_j}\frac{\partial}{\partial q_j} - \frac{\partial H}{\partial q_j}\frac{\partial}{\partial p_j}\right); \qquad \sigma_\rho(\mathcal{L}) = \frac{1}{\rho}||H||_\rho.$$

Consequently, for Hamilton's function of the class $\mathcal{A}_\rho$ equation (2.1) for the classical probability distribution on a non-Archimedean phase space is solvable in the space $\mathcal{A}'_\rho$ of distributions on the interval $|t|_K < \rho/b||H||_\rho$. Furthermore, $||\nabla H||_\rho \le \rho^{-1}||H||_\rho$. Let H be a polynomial, $\deg H = s$. Then $||H||_\rho \le \rho^s(\max|H_{\alpha\beta}|_K)$. Consequently, $\kappa = \overline{\lim}\sigma_\rho(\mathcal{L}) < \infty$ if $s \le 2$. Thus, for Hamilton's quadratic functions Liouville's equation (2.1) is solvable in the space $\mathcal{A}'$ of distributions.

Let us now study Liouville's equation (2.2) in $L_2(K^n,\nu)$ ($\mathcal{A}_\rho \subset L_2$, $\rho > \sqrt{\frac{b}{|2|_K}}$). We shall consider (2.2) on the subspace $\mathcal{A}_\rho$ of the space L_2:

$$\sigma_\rho(\mathcal{L}+u) = \max(\rho^{-1}||\nabla H||_\rho; ||H||_\rho) \le ||H||_\rho.$$

We find that Liouville's equation (2.2) for the density of probabilities $\mathrm{P}(t,\Omega)$ is solvable on the interval $|t|_K < 1/b||H||_\rho$.

Using the standard formula for the transition from Schrödinger's picture to that of Heisenberg and relation (3.3), we find that for Hamilton's function $H(q,p) = |p|^2/2 + v(q)$, $v \in \mathcal{A}_\rho$, Heisenberg's equation (1.5) is solvable in the space of the linear continuous operators $\mathcal{L}(\mathcal{A}_\rho, \mathcal{A}_\rho)$ on the interval

$$|t|_K < \min\left[\frac{|h|_K}{b||v||_\rho\sqrt{|\tau|_K}}, \frac{\rho^2|2\sqrt{\tau}|_K}{b|h|_K}\right].$$

We set

$$\mathcal{L}_{quantum} = \sum_{n=0}^{\infty}\frac{1}{n!}\left(\frac{h}{\sqrt{\tau}}\right)^{n-1}\left(\left(\frac{\partial^n H}{\partial p^n}, \frac{\partial^n}{\partial q^n}\right) - \left(\frac{\partial^n H}{\partial q^n}, \frac{\partial^n}{\partial p^n}\right)\right);$$

$$\sigma_\rho(\mathcal{L}_{quantum}) \le \max_n\left[\left|\frac{h}{\sqrt{\tau}}\right|_K^{n-1}\rho^{-2n}b^n\right]||H||_\rho.$$

By virtue of Theorem 3.1, in the space $\mathcal{A}_\rho$ the solution of equation (1.6) for the symbols of PDO exists on the interval

$$|t|_K < \min_n[\rho^{2n}\left|\frac{\sqrt{\tau}}{h}\right|_K^{n-1}/b^{n+1}||H||_\rho].$$

5. Two Measurement Processes: a Scale with an Infinite Decrease of 1 and a Scale with an Infinite Increase of 1

Every process of measurement presupposes that we are given a unit of measurement 1 and a coefficient m of increase and decrease of 1, i.e., a scale $S(1,m)$. Only quantities of the form

$$x = \frac{a_{-k}}{m^k} + \ldots + a_0 + a_1 m + \ldots + a_n m^n, \tag{5.1}$$

where $a_j = 0, 1, \ldots, m-1$, can be actually measured with the use of this scale. It is natural to call these quantities $S(1,m)$-measurable.

We have two equal possibilities.

1. We suppose that the process of decrease of unity m times can be infinite, i.e.,

$$x = \ldots + \frac{a_{-k}}{m^k} + \ldots + a_0 + a_1 m + \ldots + a_n m^n. \tag{5.2}$$

2. We suppose that the process of increase of unity m times can be infinite, i.e.,

$$x = \frac{a_{-k}}{m^k} + \ldots + a_0 + a_1 m + \ldots . \tag{5.3}$$

In the first case real numbers appear (5.2) and in the second case m-adic numbers appear (and, in particular, if $m = p$ is prime, then p-adic numbers appear). The scale $S(1,m)$ is extended either to an m-ary scale $\overleftarrow{S}(1,m)$ or to a m-adic scale $\overrightarrow{S}(1,m)$. In the classical real physics only abstractions of form (5.2) were admissible whereas in the new non-Archimedean physics abstractions of form (5.2) and abstractions of form (5.3) are admissible.

Expression (5.2) and expression (5.3) are only symbols which denote infinite algorithms of measurement. We can say that the quantities described by the infinite algorithms (5.2) and (5.3) are commensurable.

The irrational numbers (5.2) and rational numbers (5.2) that can be represented as periodic m-ary fractions are $S(1,m)$-nonmeasurable. For instance, in the decimal scale the irrational number

$$\sqrt{3} = 1 + \frac{7}{10} + \ldots \tag{5.4}$$

and the rational number

$$\frac{1}{3} = \frac{3}{10} + \frac{3}{10^2} + \ldots \tag{5.5}$$

are both nonmeasurable. We cannot obtain either $\sqrt{3}$ or $1/3$ as a result of measurements in the decimal scale but can only get their approximations as fractions (5.1), $m = 10$, i.e., for the decimal scale both

$\sqrt{3}$ and $1/3$ are only symbols used to denote the infinite algorithms of measurement (5.4), (5.5).

The same is true of m-adic measurements. For instance, in the 2-adic scale

$$\frac{1}{3} = 1 + 2 + 2^3 + 2^5 + \dots \tag{5.6}$$

Thus the infinity of the algorithms of measurement with an infinite decrease of unity m times (5.2) and an infinite increase of unity m times (5.3) does not reflect any special properties of quantities described by these algorithms but only depends on the choice of the scales $\overleftarrow{S}(1, m)$ and $\overrightarrow{S}(1, m)$.

The m-adic number (5.3) describes measurement process in which at every step the unit of measurement 1 is increased m times. Suppose that algorithm (5.3) is infinite. Then the m-adic number (5.3) contains an infinite number of measurement units 1 and can be interpreted as a quantity infinitely large as compared to the unit of measurement 1.

In this case a rather unusual property of infinitely large quantities is realized, namely, quantities which are finite with respect to the real metric (say, (5.6)), can be realized as infinitely large quantities in m-adic processes of measurement. We have here a dependence on the structure of the process of measurement.

6. Non-Archimedean Cosmology

We shall try here to apply m-adic numbers to cosmology. In the suggested m-adic cosmological model the m-adic numbers describe infinite distances, time intervals and masses appearing in cosmological scales.

6.1. The Model of the m-adic Space–Time in the m-adic Cosmology

The m-adic space–time possesses a number of interesting properties. For example, a shift (5.6) is possible by a distance equal to $1/3$ of the length unit 1 in an infinite number of steps of the unit length 1. Similarly, having reckoned an infinite number (5.6) of time intervals of unit length $\Delta t = 1$ from the initial moment $t_0 = 0$, we come back to the time moment $t = 1/3$.

Thus, in the m-adic model being considered the order relation in the space–time disappears on cosmological scales. A part of the measurement unit 1 is realized as an infinite number of units 1. In the m-adic model the cosmological space–time has a non-Archimedean structure.

Furthermore, for the scale $\vec{S}\ (1, 2)$ we have

$$-1 = 1 + 2 + 2^2 + \ldots + 2^n + \ldots . \tag{6.1}$$

Consequently, in the m-adic model the distinction between the directions in space and that between the past and the future disappear. By virtue of (6.1), having reckoned an infinite number (6.1) of intervals $\Delta t = 1$ from the initial time moment $t_0 = 0$, we return to the past by one time interval. We can similarly interpret the simple generalization of relation (6.1)

$$-2^n = 2^n + 2^{n+1} + \ldots . \tag{6.2}$$

Next,

$$0 = 1 + (1 + 2 + 2^2 + \ldots + 2^n + \ldots). \tag{6.3}$$

Consequently, in the m-adic model being considered the cosmological space–time has the structure of a discrete torus of an infinite radius.

During the cosmological time $T =$ (6.3) the m-adic Universe evolves to the initial moment $t_0 = 0$.

6.2. Infinite, Negative and Zero Masses in Cosmology

Infinitely large masses (5.3) are realized in the toroidal m-adic space–time and some configurations with an infinite mass may produce the null effect of a zero (say, (6.3)) or negative (say, (6.1)–(6.2)) mass.

Remark 6.1. *In the theory of the relativistic boson string the infinite mass of the ground state is also regulated by a negative number, i.e.,*

$$M_0^2 = 1 + 2 + \ldots + n + \ldots = -\frac{1}{12}. \tag{6.4}$$

It stands to reason that relation (6.4) is senseless in the field of real numbers R since the real algorithms (5.2) realize an infinite decrease of the unit of measurement 1 rather than an infinite increase of it as in (6.4). Algorithm (6.4) is closer to the m-adic algorithms (5.3).

A p-string (p is prime) is a string whose frequencies $\nu_n = p^n$, $n = 0, 1, \ldots,$

$$x_p(\sigma) = x_p^0 + \sum_{n=0}^{\infty} x_p^n \cos p^n \sigma, \qquad \sigma \in [0, \pi].$$

The square of mass for the ground state of a p-string is

$$M_0^2 = 1 + p + p^2 + \ldots + p^n + \ldots = \frac{1}{1-p}.$$

In particular, in the field of 13-adic numbers $\mathbb{Q}_{13}$ the series

$$M_0^2 = 1 + 13 + 13^2 + \ldots + 13^n + \ldots = \frac{1}{1-13} = -\frac{1}{12}$$

converges.

Consequently, the square of mass for the ground state of the p-string for $p = 13$ coincides with the square of mass for the ground state of an ordinary boson string.

It should be emphasized that in this case $2p = 26$ is a critical dimension of a boson string.

6.3. The Disappearance of Singularity for the Density of Mass

The singularity of the density of mass (its infinite increase) leads to a zero density of mass in accordance with algorithm (6.3).

Similarly, an infinite increase in the strength of a gravitational or electromagnetic field (6.3) leads to its disappearance.

7. Microcosm and the Non-Archimedean Structure of the Real Minkowski Space–Time

Here I suggest an m-adic model of the microcosm and introduce a non-Archimedean m-adic structure in the Minkowski space–time. The m-adic model of the microcosm can be described as follows.

The quantities describing events occurring in the microcosm are infinitesimal as compared to those of the macrocosm. For an observer in the coordinate system of the microcosm the quantities characteristic of the macrocosm are realized by infinitely large m-adic numbers.

For an observer in the coordinate system of the macrocosm the representation of a physical quantity (a coordinate, momentum, mass energy, charge, field strength) as (5.3) is a representation in the form of an infinite number (5.3) of inputs of microscopic quantities.

In its essence, the macrophysics is a cosmological theory of the microcosm.

The real Minkowski space–time also has a non-Archimedean structure. We can consider it to be Archimedean only approximately (neglecting the microstructure of physical quantities).

For instance, suppose that $L = 1$ is a unit of measurement in macroscales and $l = 1$ is a unit of measurement in microscales. If we do not take into account the microstructure of the quantity L, then we can consider the Archimedean relation

$$L > L/n, \qquad n = 2, 3, \ldots. \tag{7.1}$$

to be valid. However, if we take the microstructure of the quantity L into account, then for instance,

$$L = \left[\frac{a_{-k}}{m^k} + \ldots + a_0 + a_1 m + a_2 m^2 + \ldots\right] l; \tag{7.2}$$

$$\frac{L}{m} = \left[\frac{a_{-k}}{m^{k+1}} + \ldots + \frac{a_0}{m} + a_1 + a_2 m + \ldots\right] l. \tag{7.3}$$

Both quantities (7.2) and (7.3) are infinitely large as compared to l and inequality (7.1) turns into a congruence of two infinities.

Suppose, for instance, that in a 3-adic scale $\vec{S}\,(1,3): \; L = 2 + 2 \cdot 3 + 2 \cdot 3^2 + \ldots$, then $L/2 = 1 + 3 + 3^2 + \ldots$, $L - L/2 = 1 + 3 + 3^2 + \ldots = 1/(1-3) = -1/2$.

All the equalities (5.1)–(6.3) also admit of a microcosm interpretation. For instance, the positive macroscopic mass (6.1), which is infinitely large as compared to the measurement unit $\Delta = 1$ of mass in the microcosm, produces an effect of a negative mass in the microcosm whereas a similar positive time interval in the macrocosm produces an effect of a turn of time in the microcosm.

Naturally, the definition of the macrocosms and microcosms is relative and depends only on the scales of the observer. From the point of view of a cosmological observer we live in a microcosm and, neglecting the scales of quantities of our cosmos, a cosmological observer supposes that the cosmological space–time is Minkowski's real time $\mathbb{R}^4$.

8. Models with an Infinitely Large Number of Particles

The quantum operators of the field $\hat{\varphi}(x)$ and the number of particles $\hat{\mathrm{N}}$ do not commute, and therefore, for a fixed value of the strength of the field $\varphi(x)$ the number of particles is not defined and can be infinite.

In the scale $\vec{S}\,(1, m)$ the infinite number of particles is described by the m-adic integer

$$x_\infty = a_0 + a_1 m + \ldots + a_n m^n + \ldots . \tag{8.1}$$

In particular, vacuum (the state in which $\hat{\mathrm{N}}\varphi_0 = 0$) is realized as the limit of states with an infinitely growing number of particles

$$x_0 = 1 + [(m-1) + (m-1)m + \ldots + (m-1)m^n + \ldots]. \tag{8.2}$$

As usual, we establish a relationship between the quantum field system for which states with an infinite number of bosons are admissible and a system of harmonic oscillators for which states with infinite energy (8.1) are admissible.

Can we compare different infinitely large quantities defined by a measurement algorithm of the type of (5.3)? As we know, the rings of the m-adic numbers $\mathbb{Q}_m$ (and, in particular, the fields of p-adic numbers $\mathbb{Q}_p$) are not ordered. However, we can introduce on $\mathbb{Q}_m$ a partial-order relation which is obviously sufficient for applications.

Suppose we are given two numbers $x = (5.3)$ and

$$y = b_{-t} m^{-t} + \ldots + b_0 + b_1 m + \ldots . \tag{8.3}$$

We define the relation $<$ according to the following law: $y < x$ if there is a number s such that $a_s > b_s$ and $a_j \geq b_j$ for $j > s$.

We denote by $\mathbf{Q}_{fin}(m)$ the set of real numbers of type (5.1).

Proposition 8.1. *On the set $\mathbf{Q}_{fin}(m)$ the relation $<$ coincides with the ordinary order relation on $\mathbf{Q}$.*

Proof. Let $x, y \in \mathbf{Q}_{fin}(m)$ and $y < x$. Then

$$x = x_{-r}m^{-r} + \ldots + x_s m^s + \ldots + x_T m^T,$$

$$y = y_{-t}m^{-t} + \ldots + y_s m^s + \ldots + y_T m^T,$$

with $y_s < x_s$ and $y_j \leq x_j$, $j > s$. The worse case is when $x = x_s m^s + \ldots + x_T m^T$ and $y = (m-1)(m^{-t} + \ldots + m^{s-1}) + (x_s - 1)m^s + \ldots + x_T m^T$. But $(m-1)(m^{-t} + \ldots + m^{s-1}) < m^s$. Consequently, $y < x$ in the sense of an ordinary order relation on $\mathbf{Q}$.

As usual, the relation $<$ generates a relation $\leq$.

Proposition 8.2. *The relation $\leq$ is a (partial) order relation.*

Proof. 1. It is obvious that $x \leq x$. 2. Let $x \leq y$ and $y \leq x$. Then there exist $s_1 : x_{s_1} > y_{s_1}$, $x_j \geq y_j$, $j > s_1$ and $s_2 : y_{s_2} > x_{s_2}$, $y_j \geq x_j$, $j > s_2$. Suppose, for instance, that $s_1 > s_2$. Then, on one hand, $y_j \geq x_j$, $j = s_1$ and, on the other hand, $x_{s_1} > y_{s_1}$. We arrive at a contradiction. It remains to suppose that all $x_j = y_j$. 3. Let $y \leq x$ and $x \leq z$. Then there exist $s_1 : y_{s_1} < x_{s_1}$, $y_j \leq x_j$, $j > s_1$, and $s_2 : x_{s_2} < z_{s_2}$, $x_j \leq z_j$, $j > s_2$. It remains to take $s = \max(s_1, s_2)$.

The relation $\leq$ is only a partial order relation. For instance, two numbers $x = m + m^3 + m^5 + \ldots$ and $y = 1 + m^2 + m^4 + \ldots$ are not compared. We say that the numbers from $\mathbf{Q}_{fin}(m)$ are finite, the numbers from $\mathbf{Q}_m \backslash \mathbf{Q}_{fin}(m)$ are infinite and every finite number is smaller than any infinite number.

Now we can compare, for instance, the numbers of particles for two states with an infinitely large number of particles.

9. The p-adic Interpretation of Tachyons

We know that in the special theory of relativity the momentum can be found from the formula

$$k = Mv/\sqrt{1 - \frac{v^2}{c^2}} = Mvc/\sqrt{c^2 - v^2}. \tag{9.1}$$

If the velocity $v \geq c$, then this formula becomes meaningless. However, as has been mentioned, in the fields of p-adic numbers, $p \equiv 1 \pmod 4$, there exists a square root of (-1). Thus formula (9.1) has sense for supersonic velocities v as well.

Let us consider,for example, the case $p = 5$. This choice of the coefficient of a measurement generates results belonging to $Q_{fin}(5)$. So these are finite fractions with respect to powers of 5. Now let mass M and velocity c has been measured with the aid of the coefficient 5. Thus, they belong to $Q_{fin}(5)$. We use the same symbols M and c to denote these rational numbers. Let us consider,for instance, the velocity $v = 5c$. Then we get two values of momentum k:

$$k_1 = 0,010323031034221444...Mc$$

$$k_2 = 0,044121413410223000 1...Mc \tag{9.2}$$

These are the algorithms,where the unit of the measurement is increasing 5 times during an infinite number of steps. Thus, they are infinite numbers from our point of view. Thus, in the framework of the suggested 5-adic interpretation, a tachyon can be regarded as a particle with an infinitely large momentum. It should be pointed out that the mass and the velocity of a particle are finite. Our p-adic interpretation of tachyons agrees with the special theory of relativity. If we remain in the world of momenta commensurable with our measurement units, the light velocity cannot be exceeded. In order to break through the light barrier, a particle must be imparted an infinitely large momentum.

From the point of view of energy, in the framework of the p-adic interpretation a tachyon is a particle with a finite mass M moving faster than light and having an infinite energy. In the case being considered, for $v = 5c$ we have

$$E_1 = 0,010323031034221444...Mc^2$$

$$E_2 = 0,044121413410223000 1...Mc^2 \quad (9.3)$$

An infinitely great energy E is needed in order to break through the light barrier.

Let us now consider another model. Let a particle have an infinitely large p-adic mass

$$M = M_0 + M_1 p + \ldots + M_s p^s + \ldots .$$

Suppose again that $p = 5$ and $v = 5c$ as before. Then, if the mass of the particle $M = 1,0220340214423...$, then its momentum is $k = c$. We have thus obtained an unexpected consequence. In the framework of the p-adic theory particles can exist which have an infinite mass but a finite momentum and move with a finite, although supersonic, velocity. Thus we can define infinitely large quantities in terms of finite characteristics. In the example considered above a particle with an infinitely large mass is characterized by two finite parameters. Measuring these parameters, we can calculate the infinitely large mass $M = M_0 + M_1 p + \ldots$. To substantiate our reasoning experimentally, we must discover a particle which has, for instance, a velocity $v = 5c$ and energy $E = c^2$. Then the mass of this particle will be an infinitely large quantity as compared to the measurement unit $l = 1$.

Possibly, our considerations will be more interesting in cosmology than in the theory of elementary particles. In cosmology we predict the existence of objects which have an infinite mass, move with a finite velocity $v > c$ and have a finite energy. We also predict the existence of objects which have a finite mass, a finite supersonic velocity and an infinite energy. In the last case it is also possible to make a step towards infinitely large physical quantities by discovering an object which has, say, a velocity $v = 5c$ and mass $M = 1$. It follows from what we obtained above that the energy of this object is an infinitely large quantity as compared to the unit of measurement.

Chapter V

Functional Integrals and the Quantization of Non-Archimedean Models with an Infinite Number of Degrees of Freedom

Models in which fields assume values in the non-Archimedean field K (or its extension Z) were introduced in [64]. The quantization with the use of a functional integral was carried out in [69]. The operator quantization based on the calculus of PDO on infinite-dimensional spaces over K was carried out in [59]. Superfield models over K were discussed in [66, 18-20]. In [65] real non-Archimedean gauge fields were considered on the basis of a real space–time and a non-Archimedean group of inner symmetries. These fields are realized in tensor algebra K and $\mathbb{R}$. In [65] (see also [62]) models were suggested in which fields assume values in the Galois extensions of non-Archimedean fields. Here the group of charge conjugation Z is replaced by the Galois group G and, correspondingly, every particle has $|G|-1$ antiparticles.

1. Lagrange's Formalism for a Non-Archimedean Scalar Boson Field

The Lebesgue distribution integral serves as the basis for the construction of a non-Archimedean quantum mechanics and the field theory. Every extension $Z = K(\sqrt{\tau})$ of the field K is associated with Minkowski's metric on the space $K^4 = (x_0, \ldots, x_3)$: $g^{00} = \tau$, $g^{11} = \ldots = g^{33} = 1$, i.e.

$$(x, y)_\tau = x^1 y^1 + x^2 y^2 + x^3 y^3 + \tau x^0 y^0.$$

Wick's rotation $z_0 = x_0/\sqrt{\tau}$ in the extension Z turns Minkowski's inner product into an Euclidean inner product.

As distinct from a real field, Minkowski's form $(\cdot,\cdot)_\tau$ and Euclidean form $(\cdot,\cdot)$ possess the same properties from the point of view of positive definiteness. Indeed [17], in the field $\mathbb{Q}_p$ every nonsingular quadratic form of four variables represents all p-adic numbers. Introducing the p-adic function $\operatorname{sign}_\tau x$, we find that the signs of $\operatorname{sign}_\tau(x,x)$ and $\operatorname{sign}_\tau(x,x)_\tau$ are not defined.

The action for the non-Archimedean scalar boson field has the form

$$S(\varphi) = \frac{\tau}{2} \int\limits_{K^4} [\partial_\mu \varphi(x) \partial_\mu \varphi(x) + m^2 \varphi^2(x)] \, \mathrm{d}x + \tau g \int\limits_{K^4} V(\varphi(x)) \, \mathrm{d}x, \quad (1.1)$$

where g, $m \in K$ and the field $\varphi(x)$ assumes values in K.

Proposition 1.1. *For models of a scalar boson field with a polynomial self-action, functional* (1.1) *is analytic on* W.

Proof. The isomorphism $J : L_2(K^n, \mathrm{d}x) \to L_2(K^n, \nu)$, transfers action (1.1) into an action

$$S(\psi) = \frac{\tau}{2} \int\limits_{K^4} [(\partial_\mu - x_\mu)\psi(x)(\partial_\mu - x_\mu)\psi(x) + m^2\psi^2(x)] \mathrm{e}^{-|x|^2} \, \mathrm{d}x +$$

$$\tau g \int\limits_{K^4} \psi^n(x) \mathrm{e}^{-n|x|^2/2} \, \mathrm{d}x$$

for the model $V(\varphi) = \varphi^n$. It remains to note the operators $\partial_\mu - x_\mu$: $\mathcal{A} \to \mathcal{A}$ are continuous, $\mathcal{A}$ is a topological algebra and the distribution $\exp\{-n|x|^2/2\}\mathrm{d}x \in \mathcal{A}'$.

For a free field action (1.1) can be written as

$$S_0(\varphi) = \frac{\tau}{2}[(G\nabla\varphi, \nabla\varphi) + m^2|\varphi|^2],$$

where G is a matrix of Minkowski's τ-metric. Consequently, the action for a free field is defined on $L_2(K^n, \mathrm{d}x)$ and for interacting fields only on the space of test functions. The same is true of the field theory over $\mathbb{R}$.

We differentiate the functional $S(\varphi)$ along $h(x) \in W$:

$$\int_{K^4} \frac{\delta S}{\delta\varphi(x)}(\varphi(x))h(x)\,\mathrm{d}x =$$

$$\tau \int_{K^4} \left(\frac{\partial\varphi}{\partial x_1}\frac{\partial h}{\partial x_1} + \ldots + \tau\frac{\partial\varphi}{\partial x_0}\frac{\partial h}{\partial x_0} + m^2\varphi h \right) \mathrm{d}x +$$

$$\tau g \int_{K^4} V'(\varphi(x))h(x)\,\mathrm{d}x.$$

Applying the formula for integration by parts, we obtain

$$\int_{K^4} \frac{\delta S}{\delta\varphi(x)}(\varphi(x))h(x)\,\mathrm{d}x =$$

$$\tau \int_{K^4} \left[-\left(\frac{\partial^2\varphi}{\partial x_1^2} + \ldots + \tau\frac{\partial^2\varphi}{\partial x_0^2} \right) + m^2\varphi(x) \right] h(x)\,\mathrm{d}x +$$

$$\tau g \int_{K^4} V'(\varphi(x))h(x)\,\mathrm{d}x.$$

Hence (the inner product (5.1), Ch. 2, divides the points of the functional space W) follows the *Euler-Lagrange equation*

$$(\Box_\tau - m^2)\varphi(x) = gV'(\varphi(x)); \qquad \Box_\tau = \Delta + \tau\partial_0^2.$$

By analogy we can introduce the action for the non-Archimedean complex scalar boson field $\varphi(x) = \varphi_1(x) + \sqrt{\tau}\varphi_2(x)$:

$$S(\varphi) = \tau \int_{K^4} (\partial_\mu\bar{\varphi}(x)\partial_\mu\varphi(x) + m^2\bar{\varphi}(x)\varphi(x))\,\mathrm{d}x + \tau g \int_{K^4} V(\varphi(x), \bar{\varphi}(x))\,\mathrm{d}x,$$

$$m, g \in K.$$

As distinct from the theory of field over $\mathbb{R}$, there exist several complex non-Archimedean scalar boson fields corresponding to different quadratic extensions of the field K. For instance, three complex scalar boson fields $\varphi(x)$ with a fixed mass m were obtained for the field $\mathbb{Q}_p$, $p \neq 2$ (and for a wide class of locally compact fields [17]). In the theory of field over $\mathbb{R}$ the complex scalar boson field $\varphi(x)$ describes charged particles with charges $+$ and $-$. Therefore it is natural to call the new characteristic for a non-Archimedean field a non-Archimedean charge.

Factoring the Gordon-Klein operator $(\Box_\tau - m^2)$, we arrive at *Dirac's non-Archimedean operators* and a collection of electrons and positrons of different *non-Archimedean* charges appears.

Considering fields in the extensions of $K(\sqrt{\tau})$, we obtain particles the signs of whose charges are described not by the field Z_2 but by the field Z_n. The formalism of this work is directly extended to the fields with values in the completion of the algebraic closure of the field K. These fields have an infinite number of non-Archimedean charges with an infinitely growing number of signs.

Remark 1.1. *In [4] the authors considered closed p-adic strings. Every one of the three extensions of the field $\mathbb{Q}_p$, $p \neq 2$, is associated with a closed string, which is a circle in Z. Having connected closed strings with a non-Archimedean gravitation, Aref'eva, Volovich and Dragovic obtained a new non-Archimedean characteristic of graviton which they called a p-adic color of the graviton.*

A non-Archimedean scalar boson field is quantized by means of the functional integral

$$\int_{\mathcal{A}'(K^n,K)} G(\varphi) \exp\{\sqrt{\tau} S(\varphi)\} \prod_{x \in K^n} \mathrm{d}\varphi(x).$$

Complex and vector fields as well as *gauge non-Archimedean fields* can be quantized by analogy.

2. The Gauss and Feynman Integrals for a Non-Archimedean Quantum Scalar Boson Field

The functional integral (Feynman path integral) for a non-Archimedean scalar boson field is considered for the action in the momentum representation

$$S_0(\varphi) = \frac{\tau}{2} \int_{K^4} [\tau(k,k)_\tau + m^2]\varphi(k)\varphi(-k)\,\mathrm{d}k, \qquad |\tau|_K = 1.$$

We shall show that when certain constraints are imposed on the mass m, the operator $\mathcal{E}(k) = [\tau(k,k)_\tau + m^2]^{-1} : W \to W'$ is continuous. We use the isomorphism $J : W \to \mathcal{A}$.

Theorem 2.1. *Let* $|m|_K > \sqrt{b/|2|_K}$. *Then* $\mathcal{E}(k) \in L_2(K^n, \nu)$.

Proof. Let us consider a one-dimensional case

$$\tilde{\mathcal{E}}_n = (2^n n!\, m^2)^{-1} \sum_{j=[(n+1)/2]} \frac{(-\tau)^j}{m^{2j}}(2j-1)\ldots(2j-n) \int_K x^{2j-n}\nu(\mathrm{d}x).$$

Consequently, $\tilde{\mathcal{E}}_{2s+1} = 0$ and

$$\mathcal{E}_{2s} = (-\tau)^s(2^{2s}(2s)!m^{2+2s})^{-1}\sum_{l=0}^{\infty}(-\tau)^l(2l+2s-1)\ldots 2l(2l-1)!!(2m^2)^{-l}.$$

Therefore

$$|\tilde{\mathcal{E}}_{2s}|_K\sqrt{|2|_K^{2s}|(2s)!|_K} \le \frac{1}{|m|_K^2}\left(\frac{b}{|m|_K^2|2|_K}\right)^s.$$

The bilinear form κ on the space of test functions W is defined by the relation

$$\kappa(f,g) = \frac{1}{\tau^2}\int_{K^4}\mathcal{E}(k)f(k)\bar{g}(k)\,\mathrm{d}k.$$

Theorem 2.2. $\kappa \in \mathcal{L}_2(W^2, Z)$.

It is sufficient to show (see [69]), that the form

$$\kappa_J(f,g) = \int_{K^4} \mathcal{E}(k) f(k)\, \bar{g}(k)\, \nu(\mathrm{d}k)/\tau^2 \in \mathcal{L}_2(\mathcal{A}^2, Z).$$

The isomorphism of the spaces $J' : \mathcal{A}' \to W'$ makes it possible to introduce the Gauss and Feynman functional integrals with respect to the distribution space W' in terms of the integrals with respect to $\mathcal{A}'$:

$$\int_{W'} G(\varphi)\gamma_{a,\sqrt{\tau}b}(\mathrm{d}\varphi) = \int_{\mathcal{A}'} G(J'\varphi)\gamma_{(J')^{-1}a,\sqrt{\tau}b(J\otimes J)}(\mathrm{d}\varphi),$$

$$\int_{W'} G(\varphi)\gamma_{a,b}(\mathrm{d}\varphi) = \int_{\mathcal{A}'} G(J'\varphi)\gamma_{(J')^{-1}a,b(J\otimes J)}(\mathrm{d}\varphi).$$

In the space W we consider a K-linear space corresponding to the scalar boson field $W_B = \{\varphi : \varphi(k) = \varphi(-k)\}$. Then the Feynman integral for the scalar boson field is defined by the relation

$$\int_{W'_B} G(\varphi)\exp\{\sqrt{\tau}S_0(\varphi-\varphi_0)\} \prod_{k\in K^4} \mathrm{d}\varphi(k) = \int_{W'_B} G(\varphi)\gamma_{\varphi_0,\sqrt{\tau}\kappa}(\mathrm{d}\varphi).$$

By analogy with Theorems 2.1 and 2.2 we can prove that the function

$$\mathcal{E}_E(k) = (|k|^2 + m^2)^{-1} \in L_2(K^n, \nu)$$

if $|m|_K > \sqrt{\frac{b}{|2|_K}}$ and, consequently, the bilinear form

$$\kappa_E(f,g) = \frac{1}{\tau}\int_{K^4} \mathcal{E}_E(k) f(k)\, \bar{g}(k)\, \mathrm{d}k \in \mathcal{L}_2(W^2, Z).$$

Therefore the Euclidean Gauss functional integral

$$\int_{W'_B} G(\varphi)\gamma_{\varphi_0,\kappa_E}(\mathrm{d}\varphi)$$

is defined for entire functions on the space W'_B.

Let us calculate the generating functional for the scalar boson field:

$$\int\limits_{W'_B} \exp\{\sqrt{\tau}S_0(\varphi) + \sqrt{\tau}(\varphi, \varphi_0)\} \prod_{k\in K^4} \mathrm{d}\varphi(k) =$$

$$\int\limits_{W'_B} \exp\left\{\frac{\tau\sqrt{\tau}}{2} \int\limits_{K^4} \frac{2}{\mathcal{E}(k)}\Big([\varphi(k) + \mathcal{E}(k)J(k)/\tau]\times\right.$$

$$\left.\times[\varphi(-k) + \mathcal{E}(-k)J(-k)/\tau] - \mathcal{E}^2(k)J(k)J(-k)/\tau^2\Big)\,\mathrm{d}k\right\} \prod_{k\in K^4} \mathrm{d}\varphi(k) =$$

$$= \int\limits_{W'_B} \exp\{\sqrt{\tau}S_0(\varphi + J)\} \prod_{k\in K^4} \mathrm{d}\varphi(k) \exp\left\{\frac{\tau\sqrt{\tau}}{2} \int\limits_{K^4} \mathcal{E}(k)J(k)J(-k)\,\mathrm{d}k\right\}.$$

Using the Parseval equality, we obtain

$$\int\limits_{W'_B} G(\varphi) \exp\{\sqrt{\tau}S_0(\varphi)\} \prod_{x\in K^4} \mathrm{d}\varphi(x) =$$

$$G\left(\frac{1}{\sqrt{\tau}}\frac{\partial}{\partial\eta}\right) \exp\left\{\frac{1}{2\sqrt{\tau}} \int\limits_{K^4} \mathcal{E}(k)\eta(k)\eta(-k)\,\mathrm{d}k\right\}\Bigg|_{\eta=0}$$

With the use of relation (2.1) for the Feynman path integral we realize the perturbation theory with the *propagator* $\mathcal{E}(k)$ according to the ordinary scheme [12], [105], [111].

3. The Bargmann–Fock Representation

As was pointed out in 2, the isomorphism $J:\ \mathcal{A}' \to W'$ makes it possible to extend the theory of Gaussian integration from the space $\mathcal{A}'(K^n)$ to the space $W'(K^n)$. On the space of entire functions $\mathcal{A}(W'^2)$ acting from W'^2 to Z we consider a canonical (complex) Gaussian distribution $\gamma(\mathrm{d}\psi\mathrm{d}\bar{\psi}) = \exp\{-h^{-1}(\psi, \bar{\psi})\}\mathrm{d}\psi\mathrm{d}\bar{\psi}$. The monomials $\psi^\alpha = \psi_0^{\alpha_0} \dots \psi_n^{\alpha_n} \dots$, $|\alpha| = \sum \alpha_j < \infty$, are orthogonal with respect to the inner product $(f, g) = \int f(\psi)\overline{g(\psi)}\gamma(d\psi d\bar{\psi})$. The completion of the space $\mathcal{A}(W')$ by the non-Archimedean norm $||f|| = \max_\alpha |f_\alpha|_K \sqrt{||\psi^\alpha|^2|_K}$ is denoted by $F_2(W')$ and is called a space of analytic functions square summable with respect to $\gamma(d\psi d\bar{\psi})$.

For the quantum field Bargmann–Fock representation the space $F_2(W')$ is chosen as the state space, the quantization is carried out with the aid of PDO with the Wick symbols which are defined by means of integrals with respect to the Gaussian distribution $\gamma(d\psi d\bar{\psi})$. For the calculus of these PDO see [59]. As usual, we introduce the operators of creation and annihilation of non-Archimedean quantum particles: $\hat{a}_j^*(f)(\psi) = \psi_j f(\psi)$, $\hat{a}_j(f)(\psi) = h\frac{\partial}{\partial \psi_j} f(\psi)$.

In [59] I also consider Segal's representation for non-Archimedean quantum fields. If $\sqrt{2} \in Z$, then the Segal and Bargmann–Fock representations for non-Archimedean fields are isomorphic and if $\sqrt{2} \notin Z$, then the question concerning isomorphism is open.

Chapter VI

The p-Adic-Valued Probability Measures

The p-adic probability theory originated in connection with the problem of the probability interpretation of physical models in which wave functions assume values in the extensions of the field of p-adic numbers [64], [72]. This problem cannot be solved in the framework of Kolmogorov's standard axiomatic probability theory [94] since real numbers are included into one of Kolmogorov's axioms.

As is known (see the remarks in [94]), when he constructed the axiomatic theory in the framework of the measure theory, Kolmogorov used the von Mises frequency theory of probability [99],[100]. Every axiom of Kolmogorov is a theorem in the frequency theory of probability (except for the condition of σ-additivity which Kolmogorov imposed additionally). When constructing the non-Archimedean theory [75], [76], [71], [67], I did the same. First I constructed the general frequency theory based on the general (topological) principle of the statistical stabilization of relative frequencies. By virtue of this principle, the statistical stabilization can be regarded not only in the ordinary real topology on the field of rational numbers $\mathbb{Q}$ (and all frequencies are rational) but also in other topologies on $\mathbb{Q}$. These topologies are known as topologies of statistical stabilization and the probabilities defined as the limits of relative frequencies belong to the completion of $\mathbb{Q}$ with respect to the stabilization topology. Of especial interest are statistical samples (collectives) for which statistical stabilization is absent in the field of real numbers but is present in one of $\mathbb{Q}_p$, see

[67],[75]. On the basis of frequency theory [67],[75] I suggest axiomatic theory (the theorems of [67], [75] are taken to be axioms) based on the theory of measures and distributions with values in non-Archimedean number fields.

1. p-Adic Frequency Theory of Probability

Let us recall main notions of R. von Mises' frequency theory of probability [99],[100]. The basis of this theory is a conception of a collective. Let us consider any experiment S and let us denote by $\Omega = \{\omega_1, ..., \omega_s\}$ the set of all possible results of this experiment (attributes,labels). We consider only finite sets Ω to simplify our investigations. A set Ω is called a label set or a set of attributes. Let us consider N realizations of S and write a result x_j after each realization. Then we get a finite sequence of observations :

$$x = (x_1, ..., x_N), x_j \in \Omega. \tag{1.1}$$

A collective is an infinite idealization of (1.1) :

$$x = (x_1, ..., x_N, ...), x_j \in \Omega, \tag{1.2}$$

for which two von Mises' principles are valid.

The first one is the principle of the statistical stabilization of relative frequency of each attribute $\omega_i \in \Omega$ in the sequence (1.2). Let us compute frequencies $\nu(\omega_i) = n_i/N$,where n_i is the number of realizations of the attribute ω_i after N first observations. The principle of the statistical stabilization of the relative frequencies says :

the frequency $\nu(\omega_i)$ approaches a limit as N approaches infinity for every label $\omega_i \in \Omega$.

This limit $\mathbf{P}(\omega_i) = \lim \nu(\omega_i)$ is said to be a probability of a label ω_i in the frequency theory of probability.

The second von Mises' principle is the principle of randomness of the sequence (1.2). But this principle generated a great number of logical problems. Modifications of von Mises' theory without this principle were proposed later [119]. We shall also develop a frequency theory based only on a generalization of the principle of the statistical stabilization. In Ch. 7 we shall try to apply to the p-adic theory of probability the Kolmogorov complexity [95],see also [140]. The Kolmogorov complexity is an alternative way to the notion of randomness and I think that it will be more useful in the p-adic case then the Mises definition.

"We will say that a collective is a mass phenomenon or a repetitive event, or simply a long sequences of observations for which there are sufficient reasons to believe that the relative frequency of the observed attribute would tend to a fixed limit if the observations were infinitely continued. This limit will be called the probability of the attribute considered within the given collective [99]."

All axioms of the Kolmogorov theory of probability are theorems in the frequency theory of probability [99],[100].

It is very important for our further considerations to note that all relative frequencies belong to the field of rational numbers and real probabilities are generated with the aid of the limiting procedure with respect to the real topology (metric).

Let us develop our theory in the same way. We also investigate infinite sequences (1.2) of observations. But a new topological principle of the statistical stabilization of relative frequencies is proposed:

the statistical stabilization of the relative frequencies $\nu(\omega_i)$ may be considered with respect to an arbitrary topology τ on $\mathbb{Q}$.

This topology is said to be a topology of the statistical stabilization. Limiting values of $\nu(\omega_i)$ are said to be τ-probabilities. These probabilities belong to the completion of $\mathbb{Q}$ with respect to the topology τ.

Remark 1.1 The topology of the statistical stabilization τ must be "sufficiently good" to define a completion.

The choice of the topology τ of the statistical stabilization is connected with the concrete probability model.

A sequence (1.2),for which the principle of the statistical stabilization of the relative frequencies for the topology τ is valid,is said to be a τ-collective.

Let us denote by the symbol $\mathbf{Q}_\tau$ the completion of the field of rational numbers $\mathbf{Q}$ with respect to the topology of the statistical stabilization τ. According to [99],[100] we use a notation of a probability distribution of a τ-collective (1.2) for the sequence of probabilities

$$\mathbf{P}_x = (\mathbf{P}(\omega_1), ..., \mathbf{P}(\omega_s)).$$

The most interesting for us is the case where real topology τ_R is not a topology of the statistical stabilization for a sequence (1.2), but another topology τ is the topology of the statistical stabilization for (1.2). In this case it would be impossible to consider (1.2) as a von Mises collective. But there is a new possibility to study (1.2) as τ-collective.

Let us denote by $U_{\mathbf{Q}}$ the subset of the rational numbers field $\mathbf{Q}$:

$$U_{\mathbf{Q}} = \{q \in \mathbf{Q} : 0 \leq q \leq 1\}.$$

And let us denote by $U_{\mathbf{Q}_\tau}$ the closure of the set $U_{\mathbf{Q}}$ in the completion $\mathbf{Q}_\tau$.

The following theorem is an evident consequence of the topological principle of the statistical stabilization:

Theorem 1.1 *Probabilities* $\mathbf{P}(\omega_i)$ *belong to the set* $U_{\mathbf{Q}_\tau}$ *for an arbitrary* τ*-collective (1.2).*

As usual let us consider an algebra $\alpha(\Omega)$ of all subsets of Ω and denote by $\mathbf{P}(A), A \in \Omega$, the sum $\sum_{\omega_i \in A} \mathbf{P}(\omega_i)$. It follows from Theorem 1.1 that a probability $\mathbf{P}(A)$ belongs to the set $U_{\mathbf{Q}_\tau}$ for every $A \in \alpha(\Omega)$.

A p-adic topology τ_p on $\mathbf{Q}$ is the most important in applications to physics.

Theorem 1.2. *The set* $U_{\mathbf{Q}_p}$ *is equal to the field* $\mathbf{Q}_p$.

Proof. Let us consider the canonical expansion of a p-adic number a:

$$a = (a_{-n} + ... + p^{n-1}a_{-1} + p^n a_0 + ...)/p^n =$$

$$lim_{N\to\infty}(a_{-n} + ... + p^{n+N}a_N)/p^n(1 + p^{n+N}) = \lim_{N\to\infty} b_N,$$

where $b_N \in U_{\mathbf{Q}}$.

We have used the expression $(1 + p^{n+N})$. It may be possible to use every sequence (v_N) of natural numbers such that $\lim v_N = 1$ and $(a_{-n} + ... + p^{n+N} a_N) < p^n v_N$.

For example,every rational number may be realized as a p-adic probability. There are such "pathological" probabilities (from the point of view of the usual theory of probability) as $\mathbf{P}(A) = 2$, $\mathbf{P}(A) = 100$, $\mathbf{P}(A) = 5/3$,$\mathbf{P}(A) = -1$; it may be possible that $\mathbf{P}(A) = i_p = \sqrt{-1}$,if $p = 1(\mathrm{mod}\, 4)$.

We develop the theory of probability based on the general (topological) principle of the stabilization of relative frequencies in the same way as [99],[100].

Theorem 1.3. *Let the completion $\mathbb{Q}_\tau$ of $\mathbb{Q}$ with respect to the topology of the statistical stabilization τ be an additive topological group. Then for every τ-collective the probability is an additive function on $\alpha(\Omega)$:*

$$\mathbf{P}(A \cup B) = \mathbf{P}(A) + \mathbf{P}(B), A \cap B = \phi.$$

Proof is the same as in [100]. We only use that $\lim(u_N + v_N) = \lim u_N + \lim v_N$ in an additive topological group.

Theorem 1.4. *The probability $\mathbf{P}(\Omega) = 1$ for every topology of the statistical stabilization τ on $\mathbb{Q}$.*

We can define a conditional probability $\mathbf{P}(A/B)$ following to [100].

Theorem 1.5. *Let $\mathbb{Q}_\tau$ be a multipicative topological group. Then for arbitrary $A, B \in \alpha(\Omega), \mathbf{P}(B) \neq 0$, we have:*

$$\mathbf{P}(A/B) = \mathbf{P}(AB)/\mathbf{P}(B).$$

We only use that $\lim u_N/v_N = \lim u_N / \lim v_N$ if $\lim v_N \neq 0$ in multiplicative topological group.

Theorem 1.6. *Let $\mathbb{Q}_\tau$ be a multiplicative topological semigroup. Then $\mathbf{P}(A \cap B) = \mathbf{P}(A)\mathbf{P}(B)$ for independent events A, B.*

2. Axiomatics on the Basis of Finite-Additive Measures

There is no problem to generalize our considerations to the case of the countable set of labels $\Omega = (\omega_1, ..., \omega_s, ...)$. But it is a hard problem to realize the frequency theory of probability in the case of an arbitrary topological set of labels (compare to [100], [119]).

But we can try to use in this general case the Kolmogorov measure theoretical approach based on our frequency theory of probability and realize the main properties of the frequency probability as the axioms of a new measure-theoretical theory of probability (using the theory of $U_{\mathbf{Q}_\tau}$-valued measures as the base of such theory). Let us consider the simplest case when $\mathbf{Q}_\tau$ is a topological field.

We propose the following definition(compare to Kolmogorov [94]):

The triple $(\Omega, \alpha(\Omega), \mathbf{P})$,where

a) Ω is an arbitrary set,

b) $\alpha(\Omega)$ is an arbitrary algebra of subsets of Ω,

c) $\mathbf{P}$ is an additive function defined on $\alpha(\Omega)$ and valued in $U_{\mathbf{Q}_\tau}$ such that $\mathbf{P}(\Omega) = 1$,

is called a generalized probability space for the topology of the statistical stabilization τ on $\mathbf{Q}$.

If $\tau = \tau_R$ is the usual real topology on $\mathbf{Q}$ then we get the standard definition of the probability space with a finite-additive probability (generalized probability space [94], [34]).

The next (the main) step in Kolmogorov's construction was the consideration of the σ-additive probabilities and the extension of probability from an algebra to a σ-algebra. This condition is the foundation of the modern theory of probability. According to this condition there is well defined theory of the integration with respect to σ-additive probability (theory of probability means).

But the Kolmogorov condition is not useful in our general case. Only discrete σ-additive measures exist in the p-adic case (see, for example, [106]).

So, σ-additive condition is very well in one case (real valued probabilities) and it is not well in another case (p-adic valued probabilities). What can we do in such situation? It would be useful to realize there (on a new higher level) the old idea of R.von Mises [99] and to study probabilities not on an arbitrary set Ω (which is not connected in any way with an experiment) but on the concrete label space Ω. For every experiment we must use its own set Ω and its own topology of the statistical stabilization τ. Every pair (Ω, τ) will generate its own condition on the finite-additive probability $\mathbf{P}$ to create a well defined theory of the integration with respect to this probability.

Some geometric analogue may be considered in connection with new probability models. The situation with the Kolmogorov axiomatics of the theory of probability is similar to the situation with the Euclidean axiomatics. The Kolmogorov theory of probability is only one model of probability. We can create non-Kolmogorov theories of probability(for example,p-adic theory of probability) in the same way as non-Euclidean geometries. Different topologies of the statistical stabilization τ generate different theories of probability.

Every non-Kolmogorov theory of probability depends on two topologies. First one is the topology of the statistical stabilization τ on $\mathbf{Q}$ and the second is the topology β on the label space Ω of an experiment S. A non-Kolmogorov theory of probability is the theory based on the pair of topological spaces:

$$((\mathbf{Q}, \tau), (\Omega, \beta)).$$

We have proposed the definition of probability space in the case of the topology τ for which $\mathbf{Q}_\tau$ is a topological field. But we could consider more general topologies τ on $\mathbf{Q}$. For example,it is possible to realize our considerations for a topology of the statistical stabilization τ with discontinuous additive operation on $\mathbf{Q}$. The probability $\mathbf{P} : \alpha(\Omega) \to U_{\mathbf{Q}_\tau}$ in this non-Kolmogorov theory of probability would be nonadditive set function. Another possibility is to consider a topology τ such that addition is continuous with respect to this topology (so the probability is a finite additive function) but multiplication is discontinuous. The usual equality $\mathbf{P}(AB) = \mathbf{P}(A)\mathbf{P}(B)$ for independent events would not be true in this non-Kolmogorov theory of probability. And there is the possibility when all algebraic operations are continuous on the field of rational numbers $\mathbf{Q}$ (so $\mathbf{Q}$ is a topological field with respect to the topology of the statistical stabilization τ). But the completion $\mathbf{Q}_\tau$ can be only the topological ring. The usual equality $\mathbf{P}(A/B) = \mathbf{P}(AB)/\mathbf{P}(B), \mathbf{P}(B) \neq 0$ would not be true in this model. For example,we can consider the ring of m-adic numbers $\mathbf{Q}_m$, where m is non prime number ($m = 4, 6, 8, ...$) or some kind of adelic topology.

The previous considerations generate the next axiomatic of the general theory of probability based on the topological principle of the statistical stabilization(for the case of $\mathbb{Q}_\tau$ is the topological field).

The probability model is the system of topological and measure theoretical objects:

$$((\mathbb{Q}, \tau), (\Omega, \beta), \alpha(\Omega), \mathbf{P}, (G.I.C))$$

where

1. τ is a topology on $\mathbb{Q}$ such that $\mathbb{Q}_\tau$ is a topological field (topology of the statistical stabilization);

2. (Ω, β) is a topological space (label space);

3. $\alpha(\Omega)$ is a subalgebra of the Borel σ-algebra of (Ω, β) (algebra of events);

4. P is a finite additive set function $\mathbf{P} : \alpha(\Omega) \to U_{\mathbb{Q}_\tau}$ such that $\mathbf{P}(A) = 1$;

5. (G.I.C.) is a "good integration condition for P ",so that sufficiently large class of continuous functions $\xi : \Omega \to U_{\mathbb{Q}_\tau}$ is integrable with respect to P.

The topology of the statistical stabilization τ and the label space (Ω, β) can be found from the properties of the concrete studying probability model S and G.I.C. is generated by τ and β.

Remark 2.1. The mathematical theory of G.I.C. for general case was studied in W. Schikhof's paper [110]. There is only one problem to connect results of [110] with our probability theory. We would be able to compute sets $U_{\mathbb{Q}_\tau}$ for each case of [110] (for real case it is a segment $[0, 1]$ and for p-adic case it is $\mathbb{Q}_p$ and I don't know anything about more general topologies).

Remark 2.2. Now we cannot propose the axiomatics for nonfield topologies τ because we cannot integrate with respect to nonadditive set functions.

3. Monna-Springer Theory of Integration with Respect to Non-Archimedean Valued Measures

The integration theory for measures on *zero-dimensional locally compact* σ-compact spaces was constructed by A. Monna and T. Springer [102] (see also [101]) and generalized by W. Schikhov [110] to the case of topological spaces which are not σ-compact. We shall briefly present the main results of this theory. You could find the proofs of these results in [102].

Everywhere X is a zero-dimensional locally compact σ-compact topological space, $\Delta(X)$ is a ring of compact open subsets of X, $C_c(X)$ is the space of continuous functions $f: X \to K$ with compact supports. On this space we consider a uniform norm

$$||f|| = \max_{x \in X} |f(x)|_K.$$

Definition 3.1. *A measure is any linear functional $\mu: C_c(X) \to K$, which satisfies the following condition: for any $A \in \Delta(X)$ there exists a constant $M_A \geq 0$ such that*

$$|\mu(f)|_K \leq M_A ||f||$$

for any $f \in C_c(X)$, $\operatorname{supp} f \subset A$.

Definition 3.2. *A measure is bounded if there is a constant $M \geq 0$ such that*

$$|\mu(f)|_K \leq M||f||, \qquad f \in C_c(X);$$

the norm of the bounded measure μ is defined as the norm of an element of the dual space $C_f'(X)$ (of the K-liner normalized space $C_c(X)$): $||\mu|| = \sup_{f \neq 0} ||f||^{-1} |\mu(f)|_K$.

For any subset $B \subset X$ we use the symbol ϕ_B to denote the characteristic function B.

If $B \in \Delta(X)$, then $\phi_B \in C_c(X)$. Every measure μ generates with a function of the sets $\mu: \Delta(X) \to K$, $\mu(A) = \mu(\phi_A)$.

Theorem 3.1. *Suppose that F is the basis of the topology X consisting of compact open subsets, $\mu : F \to K$ satisfies the following conditions:*

(1) Let $A_1, \ldots, A_k \in F$, $A_i \cap A_j = \phi$, $i \neq j$, and $A = \bigcup_{j=1}^{k} A_i \in F$, and then

$$\mu(A) = \sum_{i=1}^{k} \mu(A_i);$$

(2) let $\{A\}$, $A \in F$, be a collection of the subsets of the fixed set $B \in \Delta(X)$, and then $\{|\mu(A)|_K\}$ is bounded in K. Then the function of sets μ is continued up to the measure on $C_c(X)$.

Conversely, every measure μ satisfies conditions (1) and (2).

For the function $\mu : F \to K$, satisfying conditions (1) and (2) the integral $\mu(f)$ is defined as the limit of Riemann's sums, i.e.,

$$S_u(f) = \sum_{i=1}^{k} f(a_i)\mu(A_i),$$

where $u = (\mathcal{U}, a_1, \ldots, a_k)$, $\mathcal{U} = (A_i)_{i=1}^{k}$ is the cover of $B = \operatorname{supp} f$, $A_i \in F$, $A_i A_j = \phi$, $i \neq j$, $a_j \in A_j$.

On the space $C_c(X)$ we introduce a real-valued function N_μ:

$$N_\mu(f) = \sup_{g \neq 0} ||g||^{-1} |\mu(fg)|_K, \qquad g \in C_c(X).$$

This is a non-Archimedean semi-norm on $C_c(X)$. Here are the properties of N_μ:

(a) $|\mu(f)|_K \leq N_\mu(f)$,

(b) $N_\mu(fg) \leq N_\mu(f)||g||$ (Hölder's inequality).

The *cover* $\mathcal{U} = (U_i)$ for X is said to be *special* if $U_i U_j = \phi$, $i \neq j$, $U_i \in \Delta(X)$. For the function $f \in C_c(X)$ we set

$$N_\mu(f, \mathcal{U}) = \sup_i (\sup_{x \in U_i} |f(x)| N_\mu(\phi_{U_i})).$$

Then

$$N_\mu(f) = \inf_{\mathcal{U}} N_\mu(f, \mathcal{U}).$$

For $x \in X$ we set $N_\mu(x) = \inf N_\mu(\phi_U)$, where $\{U\}$ is a system of compact open neighborhood of x.

Theorem 3.2. *Let $f \in C_c(X)$. Then*

$$N_\mu(f) = \sup_{x \in X} |f(x)|_K N_\mu(x). \tag{3.1}$$

For any $\alpha > 0$ we set

$$X_\alpha = \{x \in X : \ N_\mu(x) \geq \alpha\}, \qquad X_+ = \cup_{\alpha>0} X_\alpha,$$

$$X_0 = \{x \in X : \ N_\mu(x) = 0\}, \qquad X = X_+ \cup X_0.$$

By (3.1.) we can define $N_\mu(f)$ for any function $f : \ X \to K$ (in particular, $N_\mu(x) = N_\mu(\phi\{x\})$.

Definition 3.3. *The function f (the set A) is of μ-negligible if $N_\mu(f) = 0$ ($N_\mu(\phi_A) = 0$).*

Proposition 3.1. *The function f is of μ-negligible $\Longleftrightarrow N_\mu(f) = 0$ for any x, $f(x) \neq 0$.*

Definition 3.4. *The functions f and g are equivalent if the function $f - g$ is of μ-negligible.*

We denote by $F(\mu)$ the space of classes of μ-equivalent functions for which $N_\mu < \infty$.

Definition 3.5. *The class $L^1(X, \mu)$ of functions integrable with respect to the measure μ is defined as the closure of the space $C_c(X)$ in $F(\mu)$ (with respect to the norm N_μ).*

Proposition 3.2. *Let $f \in L^1(X, \mu)$. Then the restriction of f to any set X_α is continuous.*

Definition 3.6. *A function $f : X \to K$ is said to be absolute continuous with respect to N_μ if*

(a)for every $\epsilon > 0$ there exists $\delta > 0$ such that if $N_\mu(\phi_U) < \delta, U \in \Delta(X)$,then $N_\mu(f\phi_U) < \epsilon$;

(b)$\lim_V N_\mu(f\phi_V) = 0$,where V is a filter of sets which are complementations of compact subsets of X.

Theorem 3.3. *A function $f : X \to K$ is integrable if and only if the following conditions hold :*

(a) the restriction of f to very $X_\alpha, \alpha > 0$, is continuous;

(b) f is absolute continuous with respect to N_μ.

Proposition 3.3. *Let $f \in L^1(X, \mu)$. Then*

$$N_\mu(f) = \sup_{g \neq 0} ||g||^{-1} |\mu(fg)|_K, \qquad g \in C_c(X).$$

Everywhere below, the symbol $\overline{A}$ is used to denote the complementation of a subset A of X, $\overline{A} = X \setminus A$.

Definition 3.7. *A function $f : X \to K$ is said to be μ-measurable if for every compact $B \subset X$ and number $\epsilon > 0$ there exists such compact $B_1 \subset B$ that $N_\mu(B \cap \overline{B_1}) < \epsilon$ and the restriction of f to B_1 is continuous.*

Proposition 3.4. *A function $f : X \to K$ is μ-measurable if and only if the restriction of f to X_α is continuous for every $\alpha > 0$.*

Definition 3.8. *A sequence $\{f_n\}$ of μ-measurable functions on X is called a Egorov-convergent if for every compact B and $\epsilon > 0$ there exists a compact subspace B_1 of B such that $N_\mu(B \cap \overline{B_1}) < \epsilon$ and $\{f_n\}$ converge uniformly on B_1.*

Proposition 3.5. *A sequence $\{f_n\}$ is Egorov-convergent iff it converges uniformly on every $X_\alpha, \alpha > 0$.*

Theorem 3.4.(Limit theorem for Monna-Springer integral)*Let $\{f_n\}$ be a Egorov-convergent sequence of integrable functions. Assume that there exists an integrable function g such that $|f_n(x)|_K \leq |g(x)|_K$ for $x \in X$. Then the limit function f is integrable and $\mu(f) = \lim_{n\to\infty} \mu(f_n)$.*

As usual, we define the product $\mu \otimes \nu$ of the measures μ on X and ν on Y, with $N_{\mu\otimes\nu}(x, y) = N_\mu(x)N_\nu(y)$.

It should be pointed out in conclusion that the Radon–Nikodym theorem in the Monna–Springer theory is not valid [102].

We shall use very often the fact (Monna-Springer [102]) that there exists a Haar measure (= translate invariant measure) on $\mathbb{Q}_p$ assumed its values in $\mathbb{Q}_q$, where $p \neq q$(compare with "no Haar measure" result discussed in Ch.1 for the case of $\mathbb{Q}_p$-valued measures on $\mathbb{Q}_p$). As usual, this measure is unique with the normalization condition $\mu(\mathbb{Z}_p) = 1$. The function $N_\mu(x)$ is equal to the constant $c = 1$ and a space L^1 coincides with C_c.

4. Measures Decreasing on the Infinity

As we have already noticed,we wish to define $\mathbb{Q}_p$-valued probability measures. We are interested in a p-adic analogue of the condition of σ-additivity which was used by A.N. Kolmogorov [94] in his system of axioms. Kolmogorov [94] noticed that this condition plays only a technical role and it would be impossible to propose any practical statistical interpretation to this condition. The same role would play an analogous condition in our theory of probability.

The condition of boundedness is the first natural condition on a measure μ. Further it will be useful the following theorem.

Theorem 4.1. *A measure μ is bounded if and only if the function N_μ is bounded on X and*

$$\|\mu\| = \sup N_\mu(x).$$

Proof. Let μ be bounded; $\{U\}$ be a system of clopen neighbourhoods of the point $x \in X$. Then

$$N_\mu(x) = \inf_U \sup_{g\neq 0} \|g\|^{-1} |\mu(g\phi_U)|_K \leq \|\mu\| \sup_{g\neq 0} \|g\|^{-1} \|g\phi_U\| \leq \|\mu\|.$$

From the other hand, for every $f \in C_c(X)$:

$$|\mu(f)|_K \leq N_\mu(f) = \sup_x |f(x)|_K N_\mu(x) \leq \sup_x N_\mu(x) \|f\|.$$

Now we are going to the next condition to define $\mathbb{Q}_p$-valued probability measure. For applications we wish to have a measure not only on the ring $\Delta\Big(X\Big)$ but also on the algebra of all clopen subsets.

Definition 4.1. *A measure μ is called decreasing (on the infinity) if for every $\epsilon > 0$ there exists $U_\epsilon \in \Delta(X)$ such that*

$$\sup_{x\in \overline{U}_\epsilon} N\mu(x) < \epsilon.$$

There and everywhere below, the symbol $\overline{A}$ is used to denote the complementation of a subset A of X, $\overline{A} = X \setminus A$.

On the basis of Theorem 4.1, we get that every decreasing measure μ is bounded. There is no problem to construct examples of bounded but not decreasing measures on $\mathbb{Q}_p$.

Example 4.1. Let $X = K = \mathbb{Q}_p, \{x_n = p^{-n}\}_{n=0}^{\infty}$ and $\mu(\{x_n\}) = 1, n = 0, 1, ...$ Then μ is bounded but it is not decreasing.

Denote the algebra of clopen subsets by symbol $\Phi(X)$.

Theorem 4.2. *Let μ be a decreasing measure. Then every subset $A \in \Phi(X)$ is summable.*

Proof. We must construct such sequence of functions $g_n \in C_c(X)$ that $N_\mu(\phi_A - g_n) \to 0$. Let $\{B_n\}$ be a sequence of subsets of $\Delta(X)$ such that $\sup_{x\in\overline{B_n}} N_\mu(x) < 1/n$. Let $C_n = A \cap B_n$. These sets are clopen. Further,

$$N_\mu(\phi_A - \phi_{C_n}) = \sup_{x\in X} |\phi_A(x) - \phi_{C_n}(x)|_K N_\mu(x) < 1/n.$$

In particular,using this theorem, we get that the space X is summable.

A decreasing measure μ can be expended to the algebra$\Phi(X)$ with preserving the properties of finite-additivity and boundedness : for every $A \in \Phi(X)$

$$\sup\{|\mu\Big(B\Big)|_K : B \subset A, B \in \Delta(X)\} < \infty.$$

To prove this fact it suffices to see that

$$|\mu\Big(B\Big)|_K \leq N_\mu(\phi_B) = \sup_x \phi_B(x) N_\mu(x) \leq \|\mu\|.$$

Note also that for every $A \in \Phi(X)$ there exists such sequence $\Big(C_n\Big)$, $C_n \in \Delta(X)$, that $\mu\Big(A\Big) = \lim_{n\to\infty} \mu\Big(C_n\Big)$.

The symbol $C_b(X)$ denotes the space of continuous bounded functions $f : X \to K$. This space is endowed with the uniform norm $\|\cdot\|$.

Proposition 4.1. *Let μ be a decreasing measure. Then to every function $f \in C_b(X)$ is corresponded an element $f \in L^1(X, \mu)$ and*

$$|\mu(f)|_K \leq N_\mu(f) \leq \|f\|\|\mu\|.$$

Proof. Let $\left(B_n\right)$ be the same as in Theorem 4.2. Then

$$N_\mu(f - f\phi_{B_n}) \leq \sup_{x \in \overline{B}_n} |f(x)|_K N_\mu(x) \leq \|f\|/n.$$

Proposition 4.2. *A measure μ is decreasing if and only if the subsets X_α are compact for every $\alpha > 0$.*

Proof. 1. The sets X_α are closed. Using the decreasing of μ we get that there exists $U_\alpha \in \Delta(X)$ such that if $x \in \overline{U}_\alpha$ then $N_\mu(x) < \alpha$. That is why $X_\alpha \subset U_\alpha$.

2. Conversely,let all $X_\alpha, \alpha > 0$, be compact. There exists a clopen neighbourhood $U(x)$ for every point $x \in X_\alpha$. The system of these neighbourhoods is an open covering of the compact subset X_α. There exists a finite sub-covering $\left(U(x_j)\right)_{j=1}^n$. We set $G_\alpha = \bigcup_{j=1}^n U(x_j)$. This set belongs to $\Delta(X)$,but

$$\sup_{x \in \overline{G}_\alpha} N_\mu(x) \leq \sup_{x \in \overline{X}_\alpha} N_\mu(x) < \alpha.$$

5. The Product of a Function and a Measure

Let μ be an arbitrary measure on $X, f \in L^1(X, \mu)$. Let us consider the functional $\nu = f\mu, \nu(g) = \mu(fg)$ for $g \in C_c(X)$. Note that

$$|\nu(g)|_K \leq N_\mu(fg) \leq N_\mu(f)\|g\|.$$

That is why ν is a bounded measure.

Theorem 5.1. *Let μ be a measure,$f \in L^1(X, \mu), \nu = f\mu$. Then*

$$N_\nu(x) = |f(x)|_K N_\mu(x). \tag{5.1}$$

Proof. On the basis of definitions of $N_\nu(x)$ and$N_\mu(x)$ we get that there exists such open-compact neighbourhood U of the point x that

$$N_\nu(\phi_U) - N_\nu(x) \leq \epsilon \tag{5.2}$$

and $N_\mu(\phi_U) - N_\mu(x) \leq \epsilon$. Using $f \in L^1$, we get that there exists $g \in C_c(X)$ such that $N_\mu(f - g) \leq \epsilon$. It is possible to consider such

U that $|g(y) - g(x)|_K \leq \epsilon$ for every $y \in U$. On the basis of (5.2) it suffices to estimate

$$|N_\mu(f\phi_U) - |f(x)|_K N_\mu(x)|.$$

At first,let us note that

$$|N_\mu(g\phi_U) - |g(x)|_K N_\mu(x)|$$
$$\leq \sup_{y \in U} |g(x) - g(y)|_K N_\mu(y) + |g(x)|_K |N_\mu(\phi_U) - N_\mu(x)|$$
$$\leq \epsilon \Big(\sup_{y \in U} N_\mu(y) + |g(x)|_K \Big) \to 0, \epsilon \to 0.$$

To conclude the proof we have

$$|N_\mu(f\phi_U) - N_\mu(g\phi_U)| \leq \epsilon$$

and

$$||g(x)|_K N_\mu(x) - |f(x)|_K N_\mu(x)| \leq \sup_x |g(x) - f(x)|_K N_\mu(x) \leq \epsilon.$$

As we know every summable function f is absolutely continuous with respect to N_μ. Using (5.2), we get:

$$\forall \epsilon > 0 \exists V \in \Delta(X) : N_\mu(f\phi_{\overline{V}}) = N_\nu(\phi_{\overline{V}}) < \epsilon.$$

This result can be formulated as

Proposition 5.1. *Let μ be a measure,$f \in L^1(X, \mu)$. Then the measure $\nu = f\mu$ is decreasing.*

6. A Formula of an Exchange of Variables in the Integral of Monna-Springer with Respect to the Decreasing Measure

Everywhere below,the symbol S denotes locally-compact complete non-Archimedean field. Further we shall consider S-valued measures and functions ($K = S$).

Theorem 6.1. *Let μ be a decreasing measure and a function $\eta \in L^1(X,\mu)$. Then the composition $f \circ \eta \in L^1(X,\mu)$.*

Proof. As $\eta \in L^1(X,\mu)$ then η is continuous on X_α. That is why $M_\alpha = \eta\Big(X_\alpha\Big)$ is compact. Denote by V_α a ball in S with the centre at the zero such that $M_\alpha \subset V_\alpha$.

Further,let $U_\alpha \in \Delta(X)$ be such that $\sup_{x\in\overline{U}_\alpha} N\mu(x) < \alpha$

As $\eta \in L^1$ then for every $\delta > 0$ exists $\eta_\delta \in C_c(X)$ such that $N_\mu(\eta - \eta_\delta) < \delta$.

Let us consider a system of functions:

$$g_{\delta\alpha}(x) = \Big[\Big(f\phi_{V_\alpha}\Big) \circ \eta_\delta\Big]\phi_{U_\alpha}$$

These functions belong to $C_c(X)$. Note that

$$g_{\delta\alpha}(x) = f(\eta_\delta(x))\phi_{V_\alpha}(\eta_\delta(x))$$

for $x \in X_\alpha \subset U_\alpha$.

We prove that $N_\mu(f \circ \eta - g_{\delta\alpha}) \to 0$ if $\delta, \alpha \to 0$.

At first,we choose α such that $\|f\|\alpha < \epsilon$. Further,

$$N_\mu(f \circ \eta - g_{\delta\alpha}) \le$$

$$\max\Big[\sup_{x\in X_\alpha} |f(\eta(x)) - g_{\delta\alpha}(x)|_S N_\mu(x); \sup_{x\in\overline{X}_\alpha} |f(\eta(x)) - g_{\delta\alpha}(x)|_S N_\mu(x)\Big]$$

$$= \max[\gamma_{\delta\alpha}, \lambda_{\delta\alpha}].$$

At first, $\lambda_{\delta\alpha} \le \|f\|\alpha < \epsilon$. Further,

$$\gamma_{\delta\alpha} = \sup_{x\in X_\alpha} |f(\eta(x)) - f(\eta_\delta(x))\phi_{V_\alpha}(\eta_\delta(x))|_S N_\mu(x)$$

$$= \sup_{x\in X_\alpha} |\Big(f\phi_{V_\alpha}\Big)\Big(\eta(x)\Big) - \Big(f\phi_{V_\alpha}\Big)\Big(\eta_\delta(x)\Big)|_S N_\mu(x).$$

Now we choose $\delta = \theta\alpha$. Then $N_\mu(\eta - \eta_\delta) \le \theta\alpha$. Thus, $|\eta(x) - \eta_\delta(x)|_S N_\mu(x) \le \theta\alpha$. But this imply that $|\eta(x) - \eta_\delta(x)|_S \le \theta$ for $x \in X_\alpha$. The function $f\phi_{V_\alpha}$ has a compact support and,consequently,it is uniformly continuous. Hence, for every $\epsilon > 0$ there exists $\theta_\epsilon > 0$ such that $|f(y_1) - f(y_2)|_S < \epsilon/\|\mu\|$. for $|y_1 - y_2|_S < \theta_\epsilon$.

To conclude the proof, it suffices to choose $\delta = \delta_\epsilon = \theta_\epsilon \alpha$.

Theorem 6.2. *Let μ be a decreasing measure,$\eta \in L^1\left(X, \mu\right)$. Then the functional $\mu : C_c(S) \to S$, defined with the aid of the equality $\mu_\eta(f) = \mu(f \circ \eta)$, is a bounded measure on $C_c(S)$.*

If measure μ is bounded but not decreasing, then it is possible that $f \circ \eta \notin L^1(X, \mu)$ for $f \in C_c(S)$.

Example 6.1. Let μ be the same measure as in Example. Let us consider the function $\eta\left(x_n\right) = y_n = p^n, \eta(x) = 0$ if $x \neq x_n$. We show that $\eta \in L^1$. For every point x_n there exists a clopen ball V_n such that these balls have an empty intersection for different points. Let us consider continuous functions with compact supports

$$\eta_N(x) = \sum_{k=1}^{N} y_k \phi_{V_k}(x).$$

Then $N_\mu(\eta - \eta_N) \to 0$. Now let f be equal to $1+y$ on the ball $U_1(0)$ and the zero outside it. Then $z_n = f(y_n) = 1 + p^n$ and $\alpha_n z_n \not\to 0, n \to \infty$

Proposition 6.1. *Let μ be a decreasing measure, $\eta \in L^1(X, \mu), f \in C_c(S)$. Then*

$$N_{\mu_\eta}(f) \leq N_\mu(f \circ \eta) \tag{6.1}$$

Proof.

$$N_{\mu_\eta}(f) = \sup_{g \neq 0, g \in C_c} \|g\|^{-1} |\mu_\eta(fg)|_S \leq \sup_x |f(\eta(x))|_S N_\mu(x)$$

This inequality may be strict.

Lemma 6.1. *If $y \notin M_\alpha = \eta\left(X_\alpha\right)$, then $N_{\mu_\eta}(y) < \alpha$.*

Proof. The set M_α is closed and, hence, for a point $y \in \overline{M}_\alpha$ there exists a neighbourhood $V_\alpha : V_\alpha \bigcap M_\alpha = \emptyset$. Further,using the equality (5.1) and the inequality (6.1) we get

$$N_{\mu_\eta}(y) = \inf_{y \in V \in \Delta(X)} N_{\mu_\eta}\left(\phi_V\right) \leq \inf_{y \in V \in \Delta(X)} N_\mu\left(\phi_V \circ \eta\right) \leq$$

$$\sup_x |\phi_{V_\alpha}(\eta(x))|_S N_\mu(x) < \alpha.$$

Proposition 6.2. *Measure μ_η is decreasing for every $\eta \in L^1(X, \mu)$.*

This proposition is the consequence of the Lemma 6.1.

It would be possible to propose more weak conditions on a function f in Theorem 6.1.

Let $\eta \in L^1$. Let us introduce the functional space $\Gamma(\eta)$ of functions $f : S \to S$ which are continuous on M_α for every $\alpha > 0$. The symbol $\Gamma_b(\eta)$ denotes the subspace of $\Gamma(\eta)$ consisting of bounded functions.

Theorem 6.3. *Let μ be a decreasing measure, $\eta \in L^1(X, \mu), f \in \Gamma_b(\eta)$. Then the composition $f \circ \eta \in L^1(X, \mu)$, the function $f \in L^1(S, \mu_\eta)$ and the following equality holds*

$$\mu_\eta(f) = \mu(f \circ \eta) \tag{6.2}$$

Proof. 1. We show that $f \circ \eta \in L^1(X, \mu)$. Using the uniform continuity of the restriction f_α of f on M_α, we have that for every $\epsilon > 0$ there exists $\delta > 0$ such that $|f(x) - f(y)|_S < \epsilon$,if $|x - y|_S < \delta, x, y \in M_\alpha$. We consider the covering $U_\delta(x), x \in M_\alpha$. There exists a finite subcovering, $\bigcup_{j=1}^n U_\delta(x_j) \supset M_\alpha$. Let

$$f_{\alpha\beta}(x) = \sum_{j=1}^n f(x_j)\phi_{U_\delta(x_j)}(x)$$

for $\epsilon = \beta$. Then $f_{\alpha\beta} \in C_c(S)$ and

$$\sup_{x \in M_\alpha} |f_{\alpha\beta}(x) - f(x)|_S < \beta.$$

Now let the functions η_δ be the same as in Theorem 6.1. We consider functions $g_{\alpha\beta\delta}(x) = f_{\alpha\beta}(\eta_\delta(x))\phi_{U_\alpha}(x)$ where the set U_α is the same as in the previous Theorem. As in Theorem 6.1, it may be proved that

$$\rho_{\alpha\beta\delta} = N_\mu(f \circ \eta - g_{\alpha\beta\delta}) \to 0, \alpha, \beta, \delta \to 0.$$

Thus, the right part of the equality (6.2) is well defined.

2. Now let us show that the function $f \in L^1\left(S, \mu_\eta\right)$. It suffices to prove that there exists a sequence of functions $\{\phi_n\}, \phi_n \in C_c(S)$, such that $N_{\mu_\eta}(f - \phi_n) \to 0$. We use Lemma to prove this :

$$N_{\mu_\eta}(f - f_{\alpha\beta}) = \sup_{y \in S} |f(y) - f_{\alpha\beta}(y)|_S N_{\mu_\eta} \leq$$

$$\max\left[\sup_{y \in \overline{M}_\alpha} |f(y)|_S N_{\mu_\eta}(y); \sup_{y \in M_\alpha} |f(y) - f_{\alpha\beta}(y)|_S N_{\mu_\eta}(y)\right]$$

$$\leq \max[\alpha\|f\|, \beta\|\mu\|] \to 0, \alpha, \beta \to 0.$$

Hence, the left side of the equality (6.2) is well defined. We shall prove that left is equal to right. Using the previous considerations, we have : $\mu_\eta(f \circ \eta) = \lim \mu_\eta(f_{\alpha\beta})$. Further, $N_\mu(f \circ \eta - f_{\alpha\beta} \circ \eta) \to 0$. Thus,

$$\mu(f \circ \eta) = \lim \mu(f_{\alpha\beta} \circ \eta) = \lim \mu_\eta(f_{\alpha\beta}).$$

Remark 6.1 It may be possible to extend these results to the case of vector-valued functions $\eta : X \to S^n, \eta = (\eta_1, ..., \eta_n), \eta_j \in L^1$ The measure μ_η will be defined on $\Phi(S^n)$.

7. p-adic Valued Probability Measures

1. Axiomatics based on decreasing measures. The theory of decreasing measures is a foundation to create the p-adic valued theory of probability on an axiomatic level.

Definition 7.1 (compare to Kolmogorov [94]). *A probability space is a triple $(\Omega, \Phi(\Omega), \mathbf{P})$, where Ω is a zero-dimensional locally compact σ-compact topological space (a sample space),$\Phi(\Omega)$ is the algebra of clopen subsets (an algebra of events), $\mathbf{P}$ is a decreasing $\mathbb{Q}_p$-valued measure on $\Phi(\Omega)$ with the condition of the normalization, $\mathbf{P}(\Omega) = 1$ (probability).*

Remark 7.1. The Kolmogorov axiomatics was realized on the basis of an arbitrary set Ω and an arbitrary algebra of its subsets. But we cannot realize such idea in our theory. We cannot consider an abstract set Ω and an abstract algebra of its subsets. Topological properties (geometry) of the sample space Ω play a great role in our considerations. Our axiomatics is quit similar to the geometric probability theories of Frechet [29] or Kramer [23] (initial probability considerations of Kolmogorov were also connected with concrete geometries on a sample space).

Let S be a locally compact non-Archimeden field (as in the previous paragraph) containing $\mathbf{Q}_p$ as its subfield.

Definition 7.2. *A function $\xi : \Omega \to S$ of the class L^1 is said to be a random variable (RV).*

We introduce a mean value of ξ with the aid of the standard definition:

$$\mathbf{M}\xi \equiv \int_\Omega \xi(\omega)\mathbf{P}(d\omega) = \mathbf{P}(\xi)$$

and moments $\mathbf{m}_k(\xi) \equiv \mathbf{M}\xi^k$ (if $\xi^k \in L^1$ then these moments are well defined),and a probability distribution $\mathbf{P}_\xi(A) = \mathbf{P}(\xi \in A)$ (it is a decreasing normalized measure on S). A vector $\xi = (\xi_1, ..., \xi_n)$,where ξ_i are RV, is called a random vector. A probability distribution $\mathbf{P}_\xi$ of a random vector ξ may be introduced in the similar way as for RV (a decreasing normalized measure on S^n). We introduce mixed moments of random vector : $\mathbf{m}_\alpha(\xi) = \mathbf{M}\xi_1^{\alpha_1}...\xi_n^{\alpha_n}$

2. Independent random variables.

Definition 7.3. *RV ξ and η are called independent RV if*

$$\mathbf{P}(\xi \in A, \eta \in B) = \mathbf{P}(\xi \in A)\mathbf{P}(\eta \in B) \tag{7.1}$$

for all $A, B \in \Delta(\Omega)$.

Lemma 7.1. *Let ξ and η be independent RV. Then the following equality holds*

$$\mathbf{M}f(\xi)g(\eta) = \mathbf{M}f(\xi)\mathbf{M}g(\eta) \tag{7.2}$$

for functions $f, g \in C_c(S)$.

Proof. Let us denote by a the expression

$$|\mathbf{M}f(\xi)g(\eta) - \mathbf{M}f(\xi)\mathbf{M}g(\eta)|_S$$

Let $f^\epsilon = \sum_{k=1}^{N_\epsilon} c_k^\epsilon \phi_{U_k^\epsilon}$. be a locally constant function where $U_k^\epsilon \in \Delta(S)$ and $\sup_x |f(x) - f^\epsilon(x)|_S < \epsilon$;g^ϵ be an analogous function for g. Then $a \leq \max[a_{\epsilon 1}, a_{\epsilon 2}]$, where

$$a_{\epsilon 1} = |\mathbf{M}f(\xi)g(\eta) - \mathbf{M}f^\epsilon(\xi)g^\epsilon(\eta)|_S$$

$$\leq \max\Big[\sup_{\omega}|f(\xi(\omega))|_S|g(\eta(\omega))-g^\epsilon(\eta(\omega))|_S N_{\mathbf{P}}(\omega),$$

$$\sup_{\omega}|g^\epsilon(\eta(\omega))|_S|f(\xi(\omega))-f^\epsilon(\xi(\omega))|_S N_{\mathbf{P}}(\omega)\Big] = \max[u_\epsilon, v_\epsilon];$$

$$u_\epsilon \leq \|f\|\|\mathbf{P}\|, v_\epsilon \leq \epsilon\|g^\epsilon\|\|\mathbf{P}\|;$$

$$a_{2\epsilon} = |\mathbf{M}f(\xi)\mathbf{M}g(\eta) - \mathbf{M}f^\epsilon(\xi)\mathbf{M}g^\epsilon(\eta)|_S$$

$$\leq \max\Big[|\mathbf{M}f(\xi)\mathbf{M}(g(\eta)-g^\epsilon(\eta))|_S, |\mathbf{M}g^\epsilon(\eta)\mathbf{M}(f(\xi)-f^\epsilon(\xi))|_S\Big]$$

$$= \max[\alpha_\epsilon, \beta_\epsilon].$$

$$\alpha_\epsilon \leq N_{\mathbf{P}}(f\circ\xi)N_{\mathbf{P}}(g\circ\eta - g^\epsilon\circ\eta)$$

$$\leq N_{\mathbf{P}}(f\circ\xi)\|\mathbf{P}\|\|g-g^\epsilon\|.$$

We can estimate β_ϵ in the same way.

Theorem 7.1. *Let ξ and η be independent RV, functions $f, g \in C_b(S)$. Then the equality (7.2) holds.*

Proof. We set $M_{\alpha,\xi} = \xi(\Omega_\alpha), M_{\alpha,\eta} = \eta(\Omega_\alpha), \alpha > 0$. The set $T_\alpha = M_{\alpha\xi}\cup M_{\alpha\eta}$ is compact.

Let a function $u \in C_b(S)$. The restriction of u on T_α is uniformly continuous. Repeating the considerations of Theorem 6.3, we construct a function $u_{\alpha\beta}\in C_c(S)$ such that $\sup_{x\in T_\alpha}|u(x)-u_{\alpha\beta}(x)|_S < \beta$.

Now let $f_{\alpha\beta}$ and $g_{\alpha\beta}$ be such functions for f and g. Further,using Lemma 1,we shall only estimate expressions:

$$\lambda_{\alpha\beta} = |\mathbf{M}(f(\xi)g(\eta) - f_{\alpha\beta}(\xi)g_{\alpha\beta}(\eta))|_S;$$

$$\mu_{\alpha\beta} = |\mathbf{M}f(\xi)\mathbf{M}g(\eta) - \mathbf{M}f_{\alpha\beta}(\xi)\mathbf{M}g_{\alpha\beta}(\eta)|_S.$$

For instance, $\lambda_{\alpha\beta}\leq\max[\lambda_{\alpha\beta1},\lambda_{\alpha\beta2}]$, where

$$\lambda_{\alpha\beta1}\leq\sup_{\omega}|f(\xi(\omega))|_S|g(\eta(\omega))-g_{\alpha\beta}(\eta(\omega))|_S N_{\mathbf{P}}(\omega)$$

$$\leq \|f\|\max\Big[\sup_{\omega\in\Omega_\alpha}|g(\eta(\omega))-g_{\alpha\beta}(\eta(\omega))|_S N_{\mathbf{P}}(\omega);$$

$$\sup_{\omega\in\overline{\Omega}_\alpha}|g(\eta(\omega))-g_{\alpha\beta}(\eta(\omega))|_S N_{\mathbf{P}}(\omega)\Big].$$

If $\omega \in \Omega_\alpha$, then $\eta(\omega) \in M_{\alpha\eta} \subset T_\alpha$ and $\xi(\omega) \in M_{\alpha\xi} \subset T_\alpha$. Hence,

$$\lambda_{\alpha\beta 1} \leq \|f\| \max[\|\mathbf{P}\|\beta, \|g\|\alpha] \to 0, \alpha, \beta \to 0.$$

We can estimate $\lambda_{\alpha\beta 2}$ in the same way.

Corollary 7.1. *Let be ξ, η be independent RV. Then the equality (7.1) holds for all clopen subsets A and B.*

Everywhere below, if a property Ξ is valid on the subset Ω_+, we say that Ξ is valid a.e. $(\operatorname{mod}\mathbf{P})$.

Let RV ξ and η be bounded a.e.. Then on the basis of Theorem 6.3 we get that ξ^k and η^k are also RV for every $k = 0, 1, ...$; on the basis of Remark 6.1 we get that products $\xi^k\eta^n$ are also RV. Hence, all moments $\mathbf{m}_{kn}(\xi\eta)$ are well defined.

Theorem 7.2. *Let RV ξ and η be bounded a.e.. These RV are independent if and only if*

$$\mathbf{m}_{kn}(\xi, \eta) = \mathbf{m}_k(\xi)\mathbf{m}_n(\eta) \tag{7.3}$$

for every $k, n = 1, 2, ...$

Proof. 1. Let RV ξ and η be independent, U be a ball in S such that it contains $\xi(\Omega_+)$ and $\eta(\Omega_+)$. Then function $f_k(y) = y^k\phi_U(y)$ and $g_n(y) = y^n\phi_U(y)$ belong to the class $C_c(S)$ and

$$\mathbf{M}f_k(\xi)g(\eta) = \mathbf{m}_{kn}(\xi, \eta), \mathbf{M}f_k(\xi) = \mathbf{m}_k(\xi), \mathbf{M}g_n(\eta) = \mathbf{m}_n(\eta).$$

2. Let the condition (7.3) holds and functions f and g belong to the class $C_c(S)$. It may be possible to consider the case when supp f, supp $g \in U$. Using the fact that the set U is compact and the Kaplansky theorem, we can conclude the proof.

Let us set $\xi^c(\omega) = \xi(\omega)\phi_{U_c(0)}(\xi(\omega)), c \in \mathbf{R}_+$, for a RV ξ as in the standard theory of probability.

Proposition 7.1. *RV ξ and η are independent iff the condition (7.3) holds for all ξ^c and η^c.*

Proposition 7.2. *RV ξ and η are independent iff $\mathbf{P}_z = \mathbf{P}_\xi \otimes \mathbf{P}_\eta$ where $z = (\xi, \eta)$.*

It is the evident corollary of Lemma 7.1.

3. Conditional mean value. Let ξ and η be RV. A conditional mean value $\mathbf{M}[\xi|\eta = y]$ is defined as a function $m(y) \in L^1(S, \mathbf{P}_\eta)$ such that

$$\int_{\{\omega\in\Omega:\eta(\omega)\in B\}} \xi(\omega)\mathbf{P}(d\omega) = \int_B m(y)\mathbf{P}_\eta(dy)$$

for every set $B \in \Delta(S)$.

This definition is correct because $\phi_B(\eta(\omega))\xi(\omega) \in L^1(\Omega, \mathbf{P})$ for all RV ξ and η. It is a corollary of the following proposition.

Proposition 7.3. *Let ξ and λ be RV and λ be bounded a.e.. Then $\xi\lambda$ is a RV.*

Proof. We show that $\psi(\omega) = \xi(\omega)\lambda(\omega) \in L^1$. Let us consider a function $g(\omega) = \xi(\omega)u(\omega)$, where $u \in C_c(\Omega)$, and show that $g(\omega)$ is a RV. For every $\epsilon > 0$ there exists a function $\alpha \in C_c(\Omega)$ such that $N_{\mathbf{P}}(\xi - \alpha) < \epsilon/\|u\|$, consequently, $N_{\mathbf{P}}(\xi u - \alpha u) < \epsilon$ with $\alpha u \in C_c(\Omega)$.

Now we use the fact that $\lambda \in L^1$: there exists a sequence $(u_n), u_n \in C_c(\Omega)$ such that $N_{\mathbf{P}}(\lambda - u_n) \to 0$. Consequently,

$$\sup_{\omega\in\Omega\alpha} |\lambda(\omega) - u_n(\omega)|_S \leq (1/\alpha) \sup_{\omega\in\Omega\alpha} |\lambda(\omega) - u_n(\omega)|_S N_{\mathbf{P}}(\omega) \to 0, n \to \infty,$$

besides of this, we can consider the case$\|u_n\| \leq \sup_{\omega\in\Omega_+} |\lambda(\omega)\| = C$.

Further,we consider a sequence of RV $(\psi_N(\omega) = \xi(\omega)u_N(\omega))$. The inequality

$$|\psi_N(\omega)|_S \leq C|\xi(\omega)|_S, \omega \in \Omega_+$$

holds and (ψ_N) converge to $\xi\lambda$ uniformly on every subset $\Omega\alpha$:

$$\sup_{\omega\in\Omega\alpha} |\xi(\omega)\lambda(\omega) - \psi_N(\omega)|_S \leq \sup_{\omega\in\Omega\alpha} |\xi(\omega)|_S \sup_{\omega\in\Omega\alpha} |\lambda(\omega) -$$

$$u_N(\omega)|_S \to 0, \quad N \to \infty.$$

Here we have used the fact that the restriction of $\lambda(\omega)$ on Ω_α is continuous. To conclude the proof we use the limit theorem for the Monna-Springer integral.

The theorem of Radon-Nikodym is not valid in the non-Archimedean case [102] . It may be possible a situation when a conditional mean value does not exist.

Everywhere below,the case of bounded a.e. RV ξ is considered. Certainly, it is only a technical condition but now we cannot develop the theory without it. We suppose also that a mean value $\mathbf{M}[\lambda(\omega)|\eta(\omega)=y]$ exists everywhere we tell about it.

Proposition 7.4. *The equality*

$$\mathbf{M}f(\eta(\omega))\xi(\omega)=\mathbf{M}f(\eta(\omega))\mathbf{M}[\xi(\omega)|\eta=\eta(\omega)] \tag{7.4}$$

holds for every function $f\in\Gamma_b(\eta)$.

Proof. Let $\xi(\Omega_+)\subset U, U\in\Delta(\Omega)$. Then

$$f(\eta(\omega))\xi(\omega)\phi_U(\xi(\omega))=f(\eta(\omega))\xi(\omega),(a.e.).$$

Further, the function $g(x,y)=f(y)x\phi_U(x)$ belongs to the class $\Gamma_b(z)$ where $z=(\xi,\eta)$. Using the formula of the exchange of variables,we get

$$I=\int_\Omega g(\xi(\omega),\eta(\omega))\mathbf{P}(d\omega)=\int_{S^2} f(y)x\phi_U(x)\mathbf{P}_z(dxdy)$$

$$=\int_{S^2} f(y)x\mathbf{P}_z(dxdy).$$

We consider the measure

$$\lambda(A)=\int_{S^2}\phi_A(y)x\mathbf{P}_z(dxdy)=\int_A m(y)\mathbf{P}_\eta(dy),$$

where $m(y)=\mathbf{M}[\xi|\eta=y]$.

Hence,$I=\int_S f(y)m(y)\mathbf{P}_\eta(dy)$

Proposition 7.5. *A conditional mean value is defined uniquely a.e.* $\operatorname{mod}\mathbf{P}_\eta$.

Proof. Let $m_j(y)=\mathbf{M}[\xi|\eta=y], j=1,2$. Then

$$N_{\mathbf{P}_\eta}(m_1-m_2)=\sup_{\phi\neq 0,\phi\in C_c(S)}\|\phi\|^{-1}|\mathbf{P}_\eta(\phi(m_1-m_2))|_S.$$

But,

$$\mathbf{P}_\eta(\phi m_1)=\int_S\phi(y)m_1(y)\mathbf{P}_\eta(dy)$$

$$= \int_{\Omega} \phi(\eta(\omega))\xi(\omega)\mathbf{P}(d\omega) = \mathbf{P}_{\eta}(\phi m_2)$$

for every $f \in \Gamma_b(\eta)$.

Proposition 7.6. *The equality*

$$\mathbf{M}[f(\eta)\xi|\eta = y] = f(y)\mathbf{M}[\xi|\eta = y] \tag{7.5}$$

holds for every function $f \in \Gamma_b(\eta)$.

Proof. Certainly, we consider this equality a.e. mod $\mathbf{P}_{\eta}$. Using the formula of the exchange of variables,we get

$$\int_{\{\omega\in\Omega:\eta(\omega)\in B\}} f(\eta(\omega))\xi(\omega)\mathbf{P}(d\omega) = \int_B \phi_B(y)f(y)m(y)\mathbf{P}_{\eta}(dy).$$

The proof is concluded with the aid of the previous proposition.

Proposition 7.7. *Let* ξ *and* η *be independent RV. Then*

$$\mathbf{M}[\xi|\eta = y] = \mathbf{M}\xi. \tag{7.6}$$

Proof. Using then boundedness of ξ (a.e.),we get

$$\int_{\{\omega\in\Omega:\eta(\omega)\in B\}} \xi(\omega)\mathbf{P}(d\omega) = \mathbf{M}\xi(\omega)\mathbf{M}\phi_B(\eta(\omega))$$

$$= \int_S \phi_B(y)\mathbf{M}\xi\mathbf{P}_{\eta}(dy).$$

Proposition 7.8. *Let* ξ *and* η *be independent RV and function* $\phi \in C_b(S^2)$. *Then*

$$\mathbf{M}[\phi(\xi,\eta)|\eta = y] = \mathbf{M}\phi(\xi, y). \tag{7.7}$$

Proof. Denote by the symbol $g_1(y)$ the left part of (7.7) and by symbol $g_2(y)$ the right one. We wish to prove that $N_{\mathbf{P}_{\eta}}(g_1 - g_2) = 0$. Let us show that the equality $\mathbf{P}_{\eta}(\psi g_1) = \mathbf{P}_{\eta}(\psi g_2)$ holds for every $\psi \in C_c(S)$. But

$$\mathbf{P}_{\eta}(\psi g_1) = \int_{\Omega} \psi(\eta(\omega))\phi(\xi(\omega), \eta(\omega))\mathbf{P}(d\omega),$$

and

$$\mathbf{P}_\eta(\psi g_2) = \int_S \psi(y) \int_\Omega \phi(\xi(\omega), y)\mathbf{P}(d\omega)\mathbf{P}_\eta(dy) =$$

$$\int_{S^2} \psi(y)\phi(x, y)\mathbf{P}_z(dxdy),$$

where $z = (\xi, \eta)$.

Denote by the symbol R_α the compact set $z(\Omega_\alpha), \alpha > 0$. Let $T_\alpha \in \Delta(S^2), R_\alpha \subset T_\alpha$; it may be possible to study the case $T_\alpha = T_\alpha^1 \times T_\alpha^2, T_\alpha^j \in \Delta(S)$. In view of Lemma 6.1, we get

$$|\mathbf{P}_z(\psi\phi) - \mathbf{P}_z(\psi\phi\phi_{T_\alpha})|_S \leq N_{\mathbf{P}_z}(\psi\phi(1 - \phi_{T_\alpha}))$$

$$\leq \|\psi\| \sup_{t \in \overline{T_\alpha}} |\phi(t)|_S N_{\mathbf{P}_z}(t) \leq \alpha\|\psi\|\|\phi\|.$$

Hence, it is possible to approximate $\mathbf{P}_\eta(\psi g_2)$ with the aid of

$$\{\mathbf{P}_z(\psi\phi\phi_{T_\alpha})\}_{\alpha>0}.$$

Further, using the theorem of Kaplansky,we get that there exists a polynomial $p_\epsilon : T_\alpha \to S$ such that

$$\sup_{t \in T_\alpha} |\phi(t) - p_\epsilon(t)|_S \leq \epsilon.$$

Consequently,

$$|\mathbf{P}_z(\psi\phi\phi_{T_\alpha}) - \mathbf{P}_z(\psi p_\epsilon\phi_{T_\alpha})|_S \leq \epsilon\|\psi\|\|\mathbf{P}_z\|.$$

Thus, it is possible to approximate $\mathbf{P}_z(\psi g_2)$ with the aid of

$$\{\mathbf{P}_z(\psi p_\epsilon\phi_{T_\alpha}\}_{\alpha,\epsilon>0}.$$

But,

$$\mathbf{P}_z(\psi p_\epsilon\phi_{T_\alpha}) = \sum_{k_1,k_2} p_{\epsilon k_1 k_2}(\mathbf{M}\psi(\eta)\eta^{k_2}\phi_{T_\alpha^2}(\eta))(\mathbf{M}\psi(\xi)\xi^{k_1}\phi_{T_\alpha^1}(\xi))$$

RV ξ, η are independent and, hence,

$$\mathbf{P}_z(\psi p_\epsilon \phi_{T_\alpha}) = \sum_{k_1,k_2} p_{\epsilon k_1 k_2} \mathbf{M}\psi(\eta)\eta^{k_2}\phi_{T^2_\alpha}(\eta)\psi(\xi)\xi^{k_1}\phi_{T^1_\alpha}(\xi)$$

$$= \mathbf{M}\psi(\eta)p_\epsilon(\xi,\eta)\phi_{T_\alpha}(\xi,\eta).$$

Concluding the proof, we show that RV $\{(\psi(\eta)p_\epsilon(\xi,\eta)\phi_{T_\alpha}(\xi,\eta)\}_{\alpha,\epsilon>0}$ approximate $\psi(\eta)\phi(\xi,\eta)$. Indeed,

$$\mathcal{N}_{\mathbf{P}}(\psi(\eta)\phi(\xi,\eta)(1-\phi_{T_\alpha}(\xi,\eta)))$$

$$\leq \|\psi\|\|\phi\|\mathcal{N}_{\mathbf{P}}(1-\phi_{T_\alpha}(\xi,\eta)) \leq \|\psi\|\|\phi\| \sup_{\omega\in\overline{\Omega_\alpha}} \mathcal{N}_{\mathbf{P}}(\omega) \leq \alpha\|\psi\|\|\phi\|.$$

and

$$\mathcal{N}_{\mathbf{P}}(\psi(\eta)\phi(\xi,\eta)\phi_{T_\alpha}(\xi,\eta) - \psi(\eta)p_\epsilon(\xi,\eta)\phi_{T_\alpha}(\xi,\eta))$$

$$\leq \|\psi\| \sup_{\omega\in\Omega} |(\psi(\xi(\omega),\eta(\omega)) - p_\epsilon(\xi(\omega),\eta(\omega)))\phi_{T_\alpha}(\xi,\eta))|_S \mathcal{N}_{\mathbf{P}}(\omega) \equiv \gamma_\epsilon.$$

We set $W_\alpha = z^{-1}(T_\alpha); \Omega_\alpha \subset W_\alpha$. Then

$$\gamma_\epsilon \leq \max\Big[\sup_{\omega\in W_\alpha} |\phi(\xi(\omega),\eta(\omega)) - p_\epsilon(\xi(\omega),\eta(\omega))|_S \mathcal{N}_{\mathbf{P}}(\omega),\ \sup_{\omega\in\overline{W_\alpha}} \mathcal{N}_{\mathbf{P}}(\omega)\Big]$$

$$\leq \max[\epsilon\|\mathbf{P}\|,\alpha] \to 0, \alpha.\epsilon \to 0.$$

Corollary 7.2. *Let ξ and η be independent RV and a set $B \in \Phi(S^2)$. Then*

$$\mathbf{M}[\phi_B(\xi,\eta)|\eta = y] = \mathbf{M}\phi_B(\xi,y).$$

4. Conditional density. Let ν be a measure on S (probably, nondecreasing or unbounded). A RV ξ has a density with respect to ν if there exists a function $f_\xi(y)$ (density) of the class $L^1(S,\nu)$ such that $\mathbf{P}(\xi(\omega) \in A) = \int_A f_\xi(y)\nu(dy), A \in \Phi(S)$, so $\mathbf{P}_\xi = f_\xi \nu$. As usual, $\int_S f_\xi(y)\nu(dy) = \mathbf{P}(\xi(\omega) \in S) = 1$. The same definition will be used for random vectors.

Let a random vector $z(\omega) = (\xi(\omega),\eta(\omega))$ has a density $f_z(x,y)$ with respect to $\nu \otimes \nu : \mathbf{P}_z = f_z \nu \otimes \nu$. Set $f_{\xi|\eta}(x|y) = f_z(x,y)/f_\eta(y)$ where we set $f_{\xi|\eta}(x|y) = 0$ if $f_\eta(y) = 0$.

Proposition 7.9. *Let a random vector $z = (\xi, \eta)$ has a density with respect to $\nu \otimes \nu$ and the set $G = \{y \in S : f_\eta(y) = 0\}$ is ν-negligible. Then there exists*

$$\mathbf{M}[\xi|\eta = y] = \int_S x f_{\xi|\eta}(x|y)\nu(dx).$$

Proof. We check up the equality

$$I_1 = \int_\Omega \xi(\omega)\phi_B(\eta(\omega))\mathbf{P}(d\omega) = \int_B m(y)\mathbf{P}_\eta(dy) = I_2.$$

Using the boundedness of $\xi(\Omega_+)$, we apply the formula of the exchange of variables

$$I_1 = \int_{S^2} x\phi_B(y)f_z(x,y)\nu \otimes \nu(dxdy) = \int_S \phi_B(y)m(y)\mathbf{P}_\eta(dy) = I_2.$$

5. Discrete random variables. The case of discrete RV $\eta = (y_n)$ is considered. Here it will be possible to get the standard formula for the conditional mean value. But the non-Archimedean structure generates a new problem. The sets $A_k = \eta^{-1}(y_k)$ may be nonsummable.

Example 7.1. Let $\Omega = Z_q$ and $\mathbf{P} = dx$ be the Haar measure on Z_q assumed its values in $\mathbb{Q}_p$. Then $N_\mathbf{P}(\omega) \equiv 1$ and $L^1(\Omega, \mathbf{P}) = C(\Omega)$. Let y_n be a sequence of p-adic numbers such that $y_i \neq y_j, i \neq j$, and there exists $\lim_{j\to\infty} y_j = y_\infty$. Denote by the symbol s_l the sphere of the radius q^{-l} and introduce a function η setting $\eta \equiv y_l$ on $s_l, l = 0, 1, 2, ...$ and $\eta(0) = y_\infty$. The sets s_l are clopen and the continuity of η is the consequence of its continuity at the zero. Hence, $\eta = (y_l)_{l=o}^\infty$ is a discrete RV. But, $A_\infty = \eta^{-1}(y_\infty) = \{0\}$ and $\phi_{A_\infty} \notin C(\Omega)$.

Proposition 7.10. *Let $\eta = (y_n)$ be a discrete RV and y_k be an isolated point of the sequence (y_n). Then the set A_k is summable.*

Proof. Let $U \in \Delta(S), y_k \in U$,but $y_j \notin U, j \neq k$. Then $\phi_U(\eta(\omega)) \equiv \phi_{A_k}(\omega)$. It suffices to use Theorem 6.1 to conclude the proof.

Remark 7.2. The main important is the case when

$$\eta(\omega) = \sum_{k=1}^\infty y_k \phi_{A_k}(\omega), \tag{7.8}$$

where $A_k \in \Phi(\Omega)$. There is no problem with summability of the sets A_k.

Let us propose more general sufficient condition. The compact set Ω_α is endowed with the topology induced from Ω.

Proposition 7.11. *A set $A \subset \Omega$ is summable if and only if the sets $A_\alpha = A \cap \Omega_\alpha, \alpha > 0$, are clopen in the topological spaces Ω_α.*

Proof. 1. Let $\phi_A \in L^1(\Omega, \mathbf{P})$. In particular,the restriction of ϕ_A on Ω_α is continuous. Let U be a ball with the center at the point $t = 1 \in S$ and $a = 0 \notin U$. Then $\phi_{A_\alpha}^{-1}(U) = A_\alpha$ is a clopen set because U is a clopen set and ϕ_{A_α} is continuous.

2. Let $A_\alpha \in \Delta(\Omega_\alpha), \alpha > 0$ (the sets Ω_α are compact and,hence, the notions of clopen and compact-open coincide). Let $V_\alpha \in \Phi(\Omega)$: $V_\alpha \cap \Omega_\alpha = A_\alpha$. On the basis of the compactness of A_α we may consider that $V_\alpha \in \Delta(\Omega)$,thus, $\phi_{V_\alpha} \in C_c(\Omega)$. But,

$$N_{\mathbf{P}}(\phi_A - \phi_{V_\alpha}) \leq \sup_{\omega \in \overline{\Omega_\alpha}} |\phi_A(\omega) - \phi_{V_\alpha}(\omega)| N_{\mathbf{P}}(\omega) \leq \alpha.$$

Example 7.2. Let $\Omega = Z_q, \mu = dx$. We consider a density f : $\Omega \to \mathbf{Q}_p$ such that the restriction of f on s_l is equal to the constant f_l,where $f_l \to 0, l \to \infty, |f_l|_p < |f_{l-1}|_p$, and $f(0) = 0$. This function is continuous so it belongs to L^1. The condition of the normalization generates an additional condition $\sum f_l \mu(s_l) = 1$. We set $\mathbf{P} = f\mu$. In view of Theorem 1,we get : $N_{\mathbf{P}}(\omega) = |f(\omega)|_p$ and $\Omega_0 = \{0\}, \Omega_+ = Z_q \setminus \{0\}$. The set $A = \Omega_+$ is not a clopen set and the conditions of Remark 1 are not valid for the function. But the sets $A_\alpha = A \cap \Omega_\alpha \in \Phi(\Omega)$ and,thus, the set A is summable.

It is easy to modify this example to get the situation when the sets A_α will not be clopen sets.

Example 7.3. Let $\Omega' = Z_q \times Z_q, \mu' = \mu \otimes \delta, \mu = dx$. Here $N_{\mu'}(x, 0) = 1$ and $N_{\mu'}(x, y) = 0, y \neq 0$. We introduce the density $\overline{f}(x, y) : \overline{f}(x, 0) = f(x)$, where f is the density of the previous example, and $\overline{f}(x, y) = 0$ if $y \neq 0$. We set $\mathbf{P}' = \overline{f}\mu'$. In this case : $\Omega'_+ = \{z = (x, 0) : x \neq 0\}, \Omega_0 = \{z = (x, y) : y \neq 0\} \cup \{z = (0, y)\}$. Let $A = \Omega_+$. Then $A_\alpha = A \cap \Omega'_\alpha = \Omega_\alpha \times \{0\}$ where Ω_α are the sets of the previous example. It is evident that $A_\alpha \notin \Phi(\Omega')$. We set

$U_\alpha = \Omega_\alpha \times Z_q \in \Phi(\Omega')$. Then $A_\alpha = U_\alpha \cap \Omega'_\alpha$. It is possible to apply Proposition 7.9.

We have investigated the question about the summability of the sets $A_k = \{\omega \in \Omega : \eta(\omega) = y_k\}$ for a discrete RV $\eta(\omega)$. Now let us consider the inverse problem. We study the case when all sets A_k are summable. We are interested in conditions on a function (7.8) to be a RV.

Proposition 7.12. *Suppose that*

$$\lim_{k\to\infty} |y_k|_S N_{\mathbf{P}}(A_k) = 0.$$

Then the function (7.8) is a RV.

Proof. The functions $\eta_N(\omega) = \sum_{k=1}^N y_k \phi_{A_K}(\omega)$ are RV for all N. Thus, there exist functions $\xi_N(\omega) \in C_c(\Omega)$ such that $N_{\mathbf{P}}(\eta_N - \xi_N) \leq \epsilon$,but $N_{\mathbf{P}}(\eta_N - \xi_N) = \sup_{k>N} |y_k|_S N_{\mathbf{P}}(A_k) \to 0, N \to \infty$.

Lemma 7.2. *Let the sets $(A_k)_{k=1}^\infty$ be summable and $\cup_{k=1}^\infty A_k = \Omega, A_i \cap A_j = \phi, i \neq j$. Then $A_k \cap \Omega_\alpha \neq \phi, \alpha > 0$, only for a finite set of numbers k.*

Proof. In view of Proposition 7.11, the system of sets $A_k \cap \Omega\alpha$ is the open covering of the compact topological space Ω_α.

Proposition 7.13. *Let λ be a RV and $\eta =$ (7.8), where the sets A_k are summable, be a discrete RV, a function $f \in C_c(S)$. Then the equality*

$$\mathbf{M}f(\eta(\omega))\lambda(\omega) = \sum_{k=1}^\infty f(y_k) \int_{A_k} \lambda(\omega)\mathbf{P}(d\omega)$$

holds.

Proof. Using Theorem 6.1, we get that $z(\omega) = f(\eta(\omega))$ is a RV. This RV is bounded and discrete. We consider RV

$$u_n(\omega) = \sum_{k=1}^\infty f(y_k)\phi_{A_k}(\omega)\lambda(\omega).$$

Then $u_n(\omega) = z(\omega)\lambda(\omega)$,if $\omega \in \cup_{k=1}^n A_k$, and $u_n(\omega) = 0$, if $\omega \notin \cup_{k=1}^n A_k$,. Thus, the inequality $|u_n(\omega)|_S \leq \|f\||\lambda(\omega)|_S$ holds. We use Lemma 7.2 and the limit theorem to conclude the proof.

Proposition 7.14. *Let η be the same as in Proposition 7.13. Then $\mathbf{P}_\eta$ is a discrete measure concentrated on the points $(y_n) = \eta(\Omega)$ and $N_{\mathbf{P}_\eta}(y_k) = |\mathbf{P}(A_k)|_S$.*

This Proposition is a consequence of the previous one.

Theorem 7.3. *Let λ be a RV,η be the same as in Proposition 7.13. Then there exists a conditional mean value $\mathbf{M}[\lambda|\eta = y]$ and it is computed with the aid of the formula:*

$$\mathbf{M}[\lambda|\eta = y_n] = (1/\mathbf{P}(A_n)) \int_{A_n} \lambda(\omega)\mathbf{P}(d\omega). \tag{7.9}$$

Proof. If $\mathbf{P}(A_n) = 0$,in view of Proposition 7.14,$N_{\mathbf{P}_\eta}(y_n) = 0$. Thus, the function $m(y)$ is well defined with the aid of right part of (7.9). We need only to check up that this function is a conditional mean value:

$$\int_{\{\omega:\eta(\omega)\in B\}} \lambda(\omega)\mathbf{P}(d\omega) = \sum_{k=1}^{\infty} \phi_B(y_k) \int_{A_k} \lambda(\omega)\mathbf{P}(d\omega)$$

$$= \sum_{k=1}^{\infty} \phi_B(y_k)m(y_k)\mathbf{P}(A_k) = \int_S \phi_B(y)m(y)\mathbf{P}_\eta(dy).$$

To conclude the proof we show that $m \in L^1(S, \mathbf{P}_\eta)$. It is equivalent to

$$\lim_{n\to\infty} m(y_n)\mathbf{P}(A_n) = \lim_{n\to\infty} \int_{A_n} \lambda(\omega)\mathbf{P}(d\omega) = 0.$$

The function $\lambda(\omega)$ is summable and,thus, it is absolutely continuous :

for every $\epsilon > 0$ there exists $\delta > 0$ such that, if $N_{\mathbf{P}}(U) < \delta, U \in \Delta(\Omega)$, then $N_{\mathbf{P}}(\lambda\phi_U) \leq \epsilon$ and,in particular,

$$|\lambda(\omega)|_S N_{\mathbf{P}}(\omega) \leq \epsilon \tag{7.10}$$

for every $\omega \in U$. Now let $\omega \notin \Omega_\delta$. Then using the compactness of Ω_δ, we get that there exists a neighbourhood $V \in \Delta(\Omega)$ of the point ω such that $V \cap \Omega_\delta = \emptyset$. But, $N_{\mathbf{P}}(\phi_V) < \delta$ implies the inequality (7.10).

In view of Lemma 7.2, there exists N such that $A_n \cap \Omega_\delta = \phi$ for every $n > N$. Consequently, (7.10) holds for every point $\omega \in A_n$. Thus,

$$|\int_{A_n} \lambda(\omega)\mathbf{P}(d\omega)|_S \leq N_{\mathbf{P}}(\lambda\phi_{A_n}) = \sup_{\omega\in A_n} |\lambda(\omega)|_S N_{\mathbf{P}}(\omega) \leq \epsilon.$$

We note that here we have not used the boundedness of λ.

8. Bernoulli p-adic Valued Probabilities on the Ring of the q-adic Integers

Let $\alpha = (\alpha_0, ..., \alpha_{q-1}) \in \mathbf{Q}_p^q$ where q and p are the prime numbers and $\alpha_0 + ... + \alpha_{q-1} = 1$. These coefficients generate the discrete measure μ_α on the set $T_q = \{0, 1, ..., q-1\}$ which assumed its values in $\mathbf{Q}_p$. We introduce an infinite sequence of such measures $\{\mu_{\alpha^n}\}_{n=o}^\infty$ for coefficient vectors α^n. Let us realize $\Omega = T_q^\infty = T_q \times ... \times T_q ...$ as the ring Z_q and consider on Ω a formal direct product of measures $\{\mu_{\alpha^n}\} : \mu_A = \otimes_{n=0}^\infty \mu_{\alpha^n}$, where $A = (\alpha^0, ..., \alpha^n, ...)$ and $\alpha^n = (\alpha_0^n, ..., \alpha_{q-1}^n)$.

Element $\omega \in \Omega$ is written in the canonical way : $\omega = \omega_0 + \omega_1 q + ... + \omega_m q^m + ... \equiv (\omega_0, ..., \omega_m, ...), \omega_j \in T_q$. Denote by the symbol S_{qn} the set of all words of the length n in the alphabet $T_q : x = (x_0, ..., x_{n-1}), x_j \in T_q$, see, for example [140]. The symbol S_q denotes the set of all finite words. Set $l(x) = n$ for $x \in S_{qn}$. There is a natural product topology on the set of sequences Ω and a topology induced from Z_q. They are equivalent. We introduce in Ω a set $B_x, x \in S_q$, as the set of all sequences in Ω beginning with x. It is a base of Ω-topology. It is denoted by $F(\Omega)$. The set B_x is a ball in Z_q of the radius $q^{-l(x)}$ with the center in every point ω beginning with x. We set for $l(x) = n+1$:

$$\mu_A(B_x) = \mu_{\alpha^0}(x_0)...\mu_{\alpha^n}(x_n) \tag{8.1}$$

Theorem 8.1. *Let all coefficients $\{\alpha_j^n\}$ belong to the ring Z_p. Then μ_A can be extended to the measure on $\Omega \equiv Z_q$.*

Proof. Let $x \in S_{q(n+1)}$. Then $B_x = \cup_{j=0}^{q-1} B_{xj}$ where the symbol xj

is used to denote the word $y = (x, j)$. Then

$$\sum_{j=0}^{q-1} \mu_A(B_{xj}) = \sum_{j=0}^{q-1} \mu_A(B_x)\mu_{\alpha^n}(j) = \mu_A(B_x).$$

In view of this equality, we can prove a finite-additivity of the measure μ_A on the base $F(\Omega)$. Further, if $B_y \subset B_x$ then $y = (x, y_{n+1}, ..., y_m)$, $m = l(y) - 1$, and ,consequently,

$$|\mu_A(B_y)|_p = |\mu_A(B_x)|_p |\mu_{\alpha^{n+1}}(y_{n+1})...\mu_{\alpha^m}(y_m)|_p \leq |\mu_A(B_x)|_p.$$

In view of Theorem 3.1, we get that the function $\mu_A : F(\Omega) \to \mathbb{Q}_p$ defined by (8.1) can be extended to a measure on $\Delta(\Omega) \equiv \Phi(\Omega)$.

Measure μ_A is said to be a $\mathbb{Q}_p$-valued **Bernoulli** A-measure. Let us consider the symmetric case :$\alpha_0 = ... = \alpha_{q-1} = 1/q$. If $A = (\alpha, ..., \alpha, ...)$, then $\mu_A(B_x) = q^{-l(x)}$. It is the Haar measure on Z_q assumed its values in $\mathbb{Q}_p$. Further Bernoulli measures will be studied as probabilities. The symbol $\mathbf{P}_{Ap}$ will be used in such case instead of μ_A. The symbol $\mathbf{P}_{A\infty}$ will denote the standard real valued Bernoulli distribution with coefficients A. (in symmetric case $\mathbf{P}_{A\infty}$ is the standard symmetric Bernoulli probability distribution,which could be realized as Lebesgue measure on the real line segment $[0, 1]$). Bernoulli probability distributions correspond to an infinite series of independent trials.

Theorem 8.2 *Let μ_A be the Bernoulli measure. Then*

$$N_{\mu_A}(a) = \inf_n |\mu_A(B_{(a)_n})|_p, \tag{8.2}$$

where $(a)_n$ is the segment of the length n of the sequence a.

Proof. By the definition $N_{\mu_A}(a) = \inf\{N_{\mu_A}(U) : a \in U \in \Delta(\Omega)\}$. Using the inequality $N_{\mu_A}(U) \leq N_{\mu_A}(V), U \subset V$, we get : $N_{\mu_A}(a) = \inf_n N_{\mu_A}(B_{(a)_n})$. Now we use the equality

$$N_{\mu_A}(B_x) = \sup\{|\mu_A(U)|_p : U \subset B_x\}.$$

Let $U = \cup_{j=1}^t U_{r_j}(c_j)$ and an intersection of different balls is the empty set. Let us consider one ball $U_r(b)$ where $r = q^{-m}, l(x) \leq m$; in this case $U_r(b) \subset B_x$ and $(b)_m = (x, b_{l(x)}, ..., b_{m-1})$. Hence,

$$|\mu_A(U_r(b))|_p = |\mu_A(B_x)|_p|\mu_{\alpha^{l(x)}}(b_{l(x)})...\mu_{\alpha^{m-1}}(b_{m-1})|_p.$$

Further,

$$|\mu_A(U)|_p = |\sum_{j=1}^{t} \mu_A(U_{r_j}(c_j)|_p \leq |\mu_A(B_x)|_p.$$

So, we have proved :

$$N_{\mu_A}(B_x) = |\mu_A(B_x)|_p = \prod_{j=0}^{l(x)-1} |\mu_{\alpha^j}(x_j)|_p = \prod_{j=0}^{l(x)-1} |\alpha^j_{x_j}|_P.$$

Consequently,

$$N_{\mu_A}(a) = \inf_n N_{\mu_A}(B_{(a)_n}) = \inf_n |\mu_A(B_{(a)_n})|_p = \inf_n \prod_{j=0}^{n-1} |\alpha^j_{a_j}|_p.$$

Corollary 8.1 *Let all coefficients $\{\alpha^n_j\}$, besides of a finite number, belong to the unit sphere in $\mathbf{Q}_p$. Then the set Ω_0 is empty for the Bernoulli measure μ_A.*

Take into account that it is possible to generalize the construction of the Bernoulli measure. At first, let us consider instead of a prime number q an arbitrary natural number $m > 1 : \alpha = (\alpha_0, ..., \alpha_{m-1}), \alpha_0 + ... + \alpha_{m-1} = 1, \alpha_j \in \mathbf{Q}_p$. Here μ_α is the discrete measure on $T_m = \{0, ..., m-1\}$. The space $\Omega = T_m^\infty$ is isomorphic to the ring Z_m of the m-adic integers. The construction of the Bernoulli measure μ_A may be realized in the same way. It is also possible to study the case when a number of measures are not assumed their values in Z_p.

Theorem 8.3. *Let measures μ_{α^n} on T_m ,besides of a finite number, are assumed their values in Z_p. The set function μ_A defined with the aid of (8.1) is extended to a measure on Z_m.*

The proof is repeating the proof of Theorem 8.1. It suffices to check up that for every B_x :

$$|\mu_A(B_x)|_p \leq \sup_n \prod_{j=o}^{n-1} |\alpha^j_{x_j}|_p < \infty.$$

Example 8.1. Let $\alpha_0 = 1/3, \alpha_1 = 2/3$ and all measures $\mu_{\alpha^n} = \mu_\alpha$ (equal distributed case). We have for the measure μ_A that the maximal negligible set Ω_0 consists of all sequences for which an infinite number of coordinates is equal to 1. The function $N_{\mu_A}(\omega) = 2^{-k}$,if $|\omega| = \sum \omega_j = k$, and Ω_+ is the set of sequences for which a finite number of coordinates is equal to 1. Both sets Ω_0 and Ω_+ are dense in Ω. Take into account that $\Omega_\alpha \notin \Delta(\Omega)$ for this measure. The set Ω_+ on which this measure is concentrated is countable : elements of Ω_+ are natural numbers in Z_2. It is not a discrete measure because one-point sets are not summable with respect to it. Let us prove this fact.

If the set $A_n = \{n\}, n = n_0 + ... + n_l 2^l$, is summable, then $A_n \cap \Omega_\alpha$ is clopen in for every $\alpha > 0$. We wish to get a contradiction. Let $n \in \Omega_\alpha$. The set Ω_α consists of such sequences ω that a number of units $|\omega| \leq k = k_\alpha$. In particular,$|n| \leq k$. It suffices to study the case $|n| < k$.

Assume that there exists a neighbourhood $B_x, x = (x_1, ..., x_t)$, in Ω such that $B_x \cap \Omega_\alpha = \{n\}$. In this case, $n_0 = x_0, ..., n_t = x_t$. We study two cases :

1. Let $t \leq l$. Here we may choose $m = n + 2^{l+1}$. Then $|m| = |n| + 1 \leq k$, hence, $m \in \Omega_\alpha$. But, $m \in B_x$.

2. Let $t > l$. Here all $x_j = 0, j > l$. We choose $m = n + 2^{t+1} \in \Omega_\alpha \cap B_x$.

Consequently, we cannot define a measure of a point of the set of natural numbers, but this set is the support of this measure.

9. Biological Models Connected with the p-adic Valued Bernoulli Distributions

Probability properties of Bernoulli measure will be studied in this section in a connection with a stochastic biological model.

Let $\mathbf{P}_{A2}$ be the measure of Example 8.1. The event of an infinite outcomes of 1 is practically impossible for this probability distribution. And it is practically trustworthy event of a finite outcomes of 1.

Now we consider on Ω the usual real valued probability (in Kolmogorov sense) $\mathbf{P}_{A\infty}$. The real and 2-adic probabilities coincides for all events B_x. It is impossible to differ real and 2-adic probabilities of all events described with the aid of a finite number of trails. The stochastic model corresponding to $\mathbf{P}_{A2}$ differs from the usual probability model (described by $\mathbf{P}_{A\infty}$) by means of the principle which forbids a realization of an infinite number of 1. There is no such principle in the usual theory of probability.

We propose the following statistical biological model described with the aid of the probability space $(\Omega = Z_2, \mathbf{P} = \mathbf{P}_{A2}, \Delta(Z_2))$.

Let white and black mouses be contained in a closed box in the proportion: $W = 1/3, B = 2/3$, and this proportion is preserved during the time evolution of this population. We study the standard statistical experiment with this box with mouses. We draw a mouse from the box and write 0 or 1 for white or black mouse. Then a mouse is returned to the box.

But we consider mouses' population with one additional property. It is supposed that a terrible epidemic of black mouses occurs practically always at a moment and all black mouses dies at a moment. It is also supposed that such epidemic occurs in a statistically distant moment. The last condition we must understand in the following way.

Let us consider a large number of boxes with mouses. The relative frequency of boxes where the epidemic has occurred is very and very small at every moment.

Statistically there is the Bernoulli scheme with the weights $\alpha = (\alpha_0 = 1/3, \alpha_1 = 2/3)$ at every time moment and the condition of the momental extinction for black mouses at a sufficiently distinct time moment. This biological model is described by 2-adic probability distribution $\mathbf{P}_{A2}$ which coincides with the usual real probability distribution (with weights α) for every finite time interval.

There is also the possibility to construct the p-adic valued probabilities $\mathbf{P}_{Ap}$ with weights $\alpha = (1/3, 2/3)$ for every $p \neq 3$ (if $p = 3$ then the Bernoulli measure is not defined and it is only the distribution). But, if $p \neq 2$,then this probabilities do not describe the condition of the extinction of black mouses (as the usual real probability). At this moment we cannot propose any statistical model described with the aid of these probabilities. Such statistical models must differ from the usual real probability model of $\mathbf{P}_{A\infty}$ because,if $p \neq 2, 3$, then the maximal negligible set Ω_0 is empty. Hence, there is no any negligible subset for such probability. In particular, it would be impossible to neglect any one-point event.

The similar situation is also in the symmetric case $\alpha = (\alpha_0 = 1/2, \alpha_1 = 1/2)$, Here the probability distribution $\mathbf{P}_{Ap}$ is well defined on $\Omega = Z_2$ for every $p \neq 2$. It is the p-adic valued Haar measure on Z_2. These probabilities coincide with the standard symmetric Bernoulli probability $\mathbf{P}_{A\infty}$ for every event depending only to a finite number of outcomes. The main difference from the real case is that Ω_0 is empty. There is also impossible to neglect to any event; in particular, any one-point event.

In usual real theory we can coincide sequences of outcomes of the type $\omega = (010...0...), \omega' = (0011...11...)$, to get the uniform probability distribution on the segment $[0, 1]$ of the real line. It would be impossible in our case.

Now we introduce more complicated biological model with three species $S0, S1, S2$ with statistical weights $\alpha = (\alpha_0 = 5/13, \alpha_1 = 6/13, \alpha_2 = 2/13)$. Here μ_α is a discrete measure on $T_3 = \{0, 1, 2\}, \Omega = T_3^\infty \equiv Z_3$. Let $\mathbf{P}_{Ap}$ be the corresponding probability on Ω. We wish to describe a statistical model with the extinction of species $S1$ and $S2$ which coincides with the corresponding real model $\mathbf{P}_{A\infty}$ for every event depending only to a finite number of outcomes. This model is described with the aid of the probability $\mathbf{P}_{A2}$. If we wish to describe a model with the extinction of the specie $S0$,we may use $\mathbf{P}_{A5}$, if $S1$, then $\mathbf{P}_{A3}$.

If $p \neq 2, 3, 5, 13$, then the probability $\mathbf{P}_{\mathbf{Ap}}$, as the real probability, does not describe the extinction of any specie.

Such statistical biological models may be useful not only for the epidemic extinction at the statistically distant time moments but their may be also useful for the models where food resources are renewal for some species and nonrenewal for other species. For instance, let us consider a very large number of populations consisting of two types of animals $B0$ and $B1$ with the constant proportion of $B0$ and $B1$ during the time evolution of these populations. But $B0$ and $B1$ will be developing in a different way. A speed of the increasing of the population $B0$ suffices to renew natural food resources and a speed of the increasing of the population of $B1$ is very high to renew its natural food resources. This degeneration of $B1$-food resources is not very quickly statistically. At every moment only very little part of all $B1$-populations has already exhausted their natural food resources.

It would be interesting to compare real and p-adic probability models in this situation. These probability models do not differ on the basis of any finite time evolution. In particular,all mean values of random variables (with rational values) depending only of a finite time interval coincide. The difference is in the "global prognosis" of the evolution. If we use the usual real theory of probability,we can tell that there is a very stable situation in the evolution of $B0$ and $B1$. There is a very small fluctuation when for a number of populations $B1$-food natural resources has been exhausted. The real probability of the event G consisting of ω for which $B1$-resources must be exhausted is equal to zero. It is "practically impossible" from the real point of view. In view of the real theory of probability, we may conclude that all is very well for the $B1$-specie.

But,in view of the p-adic theory, we get another conclusion. The set G is not negligible to the p-adic Bernoulli measure. Hence, the situation is not very well for $B1$-specie. There is no problem to introduce such $\mathbf{P}_{A_p}$ that the set $\overline{G}$ is negligible. The situation is not well for $B1$ in this case.

10. A Limit Theorem for Sums of Independent Random Variables

As usual, a sequence $\{\mu_n\}$ of probability distributions is said to be weak converge to $\mu, \mu_n \overset{w}{\to} \mu$, if $\mu_n(f) \to \mu(f)$ for every function $f \in C_b$. A sequence of RV $\{\xi_n\}$ is said to be weak converge to ξ if $\mathbf{P}_{\xi_n} \overset{w}{\to} \mathbf{P}_{\xi}$.

Theorem 10.1. *Let* $\{\xi_{nj}\}_{n=1,j=1}^{\infty,n}$ *be a sequence of RV which have the moments of all orders and the estimate*

$$|\mathbf{m}_{\xi_{nj}}(k)|_S \leq d_n^k, k = 1, 2, ...,$$

holds for these moments. Let $\{c_n\}_{n=1}^{\infty}$ *be a sequence of nonzero numbers and the condition* $d_n/|c_n|_S \to 0, n \to \infty,$ *holds. If the normalized sums* $T_n(\omega) = S_n(\omega)/c_n$, *where* $S_n(\omega) = \xi_{n1}(\omega) + ... + \xi_{nn}(\omega)$, *are bounded a.e. uniformly with respect to* n,*then* $T_n \xrightarrow{w} 0, n \to \infty$.

Proof Let $U \in \Delta(S), \cup_n T_n(\Omega_+) \subset U$. Using the Kaplansky theorem, we get that for every $f \in C_b(S)$ there exists a polynomial $p_\epsilon : S \to S$ such that $\sup_{x \in U} |p_\epsilon(x) - f(x)|_S \leq \epsilon/||\mathbf{P}||$. Further,

$$|\mathbf{M}f(T_n(\omega)) - \mathbf{M}p_\epsilon(T_n(\omega))|_S \leq N_\mathbf{P}(f(T_n) - p_\epsilon(T_n))$$

$$= \sup_{\omega \in \Omega_+} |f(T_n(\omega)) - p_\epsilon(T_n(\omega))|_S N_\mathbf{P}(\omega) \leq \epsilon.$$

Now we estimate the moments of the normalized sums:

$$|\mathbf{m}_{T_n}(k)|_S = |(k!/c_n^k) \sum_{\alpha_1+...+\alpha_n=k} \mathbf{m}_{\xi_{n1}}(\alpha_1)...\mathbf{m}_{\xi_{nn}}(\alpha_n)/\alpha_1!...\alpha_n!|_S.$$

The numbers $k!/\alpha_1!...\alpha_n!$ are natural. Consequently,

$$|\mathbf{m}_{T_n}(k)|_S \leq |c_n|_S^{-k} \max_{\alpha_1+...+\alpha_n=k} |\mathbf{m}_{\xi_{n1}}(\alpha_1)|_S...|\mathbf{m}_{\xi_{nn}}(\alpha_n)|_S \leq (d_n/|c_n|_S)^k.$$

Hence,$\mathbf{M}p_\epsilon(T_n(\omega)) \to 0, n \to \infty$.

Example 10.1. Let $\{\xi_n\}$ be a sequence of independent equal distributed RV with the estimate for moments $|\mathbf{m}_\xi(k)|_S \leq d^k, d > 0$. In this case, if $T_n(\omega)$ are bounded a.e. uniformly with respect to n, then $T_n \xrightarrow{w} 0$ for every sequence $\{c_n\}$ such that $|c_n|_S \to \infty, c_n \neq 0$. In particular,it is valid if

$$|\xi_n(\omega)|_S \leq |c_n|_S, a.e. \tag{10.1}$$

Here,if $|\xi_n(\omega)|_S = o(|c_n|_S)$, a.e., then $T_n(\omega) \to 0$ a.e.. But,generally, (10.1) does not imply the convergence a.e..

Example 10.2. Let us consider the probability space for the A-measure Bernoulli $\mathbf{P}_{Ap}$ with $\alpha = (\alpha_0 = 1, \alpha_1 = a, \alpha_2 = -a)$ where $|a|_p = 1$. On the basis of Theorem 8.2, we get $N_{\mathbf{P}_{Ap}} \equiv 1, \Omega_0 = \phi, \Omega_+ = \Omega$. Hence, $L^1(\Omega, \mathbf{P}_{Ap}) = C(\Omega)$ and all RV are continuous maps. Let us introduce the sequence of RV $\xi_n(\omega) = 0$,if $\omega_n = 0$, and $\xi_n(\omega) = \gamma_n$, if $\omega_n = 1, 2$, where constants $\gamma_n \in \mathbf{Q}_p$. These RV are independent and equally distributed:

$$\mathbf{P}(\omega \in \Omega : \xi_n(\omega) = 0) = \mathbf{P}(\cup_{l(x) \leq n} B_{x0}) = \sum_{l(x) \leq n} \mathbf{P}(B_{x0}) = \mathbf{P}(\Omega) = 1$$

and

$$\mathbf{P}(\omega \in \Omega : \xi_n(\omega) = \gamma_n) = \mathbf{P}\Big(\Big[\cup_{l(x) \leq n} B_{x1}\Big] \cup \Big[\cup_{l(x) \leq n} B_{x2}\Big]\Big)$$

$$= \sum_{l(x) \leq n} \mathbf{P}(B_{x1}) + \sum_{l(x) \leq n} \mathbf{P}(B_{x2}) = \sum_{l(x) \leq n} \mathbf{P}(B_x)a - \sum_{l(x) \leq n} \mathbf{P}(B_x)a = 0.$$

Consequently,$\xi_n(\omega)$ are discrete RV which assume the value 0 with the probability 1.

We consider the sums $S_n(\omega) = \xi_0(\omega) + \ldots + \xi_n(\omega)$ and normalized sums $T_n(\omega) = S_n(\omega)/c_n$ where $\{c_n\}$ is a sequence of p-adic numbers, $|c_n|_p \to \infty$, $n \to \infty$. If the estimate $|\gamma_n|_p \leq |c_n|_p$ holds, then normalized sums are bounded. In view of the limit theorem, we get that $T_n \xrightarrow{w} 0, n \to \infty$. But there is no convergence a.e.. For example,let us consider such constants c_n that $|c_n|_p < |c_{n+1}|_p$ and let $\gamma_n = c_n$. Then,if $\omega = (1, 1, \ldots, 1, \ldots)$, we have $\xi_n(\omega) = c_n$ and $|S_n(\omega)|_p = |c_n|_p$, hence, $T_n(\omega) \not\to 0, n \to \infty$.

11. Discrete Probabilities. A p-adic Interpretation of St. Petersburg Paradox

We consider a countable set of outcomes $\Omega = (\omega_1, \ldots, \omega_n, \ldots)$. The sequence $x = (1.2)$ is called a p-adic collective if, for every label ω_j, there is stabilization in $\mathbf{Q}_p$ and $\sum \alpha_j = 1$, $\alpha_j = \mathbf{P}(\omega_j)$. Such a collective is associated with a discrete countable additive measure on the σ-algebra of all the subsets of Ω. The examples of discrete probability distributions are: (1) $\{\alpha_n = (1-p)p^n\}_{n=0}^{\infty}$ for $\mathbf{Q}_p$, (2) $\{\alpha_n = -n\, n!\}_{n=1}^{\infty}$ for any $\mathbf{Q}_p$, (3) $\{\alpha_n = n^2(n+1)!/2\}_{n=1}^{\infty}$ for any $\mathbf{Q}_p$.

A random variable is any mapping $\xi : \Omega \to \mathbf{Q}_p$, the mean $M\xi = \sum_j \xi(\omega_j)\alpha_j$, a dispersion $D\xi = M(\xi - M\xi)^2$. In our theory dispersion can also be a negative number. For instance, suppose that $\Omega = (\omega_{-1}, \omega_0, \omega_1, \ldots)$, $\alpha_{-1} = p/(p-1)$, $\alpha_n = p^n$, $n \geq 0$, and then $D\xi = 1/(1-p^3) - (1/(1-p^2))^2$ for $\xi(\omega_{-1}) = 0$, $\xi(\omega_n) = p^n$, $n \geq 0$.

Let $\mathcal{D} = (A_1, \ldots, A_n, \ldots)$, $\mathbf{P}(A_j) \neq 0$, be a complete group of disjoint events. The countable additivity of $\mathbf{P}$ yields a formula for the total probability as well as formulas for multiplication of probabilities:

$$\mathbf{P}(AB) = \mathbf{P}(B|A)\mathbf{P}(A), \quad \mathbf{P}(A) \neq 0$$

and

$$\mathbf{P}(A_1 \ldots A_n) = \mathbf{P}(A_1)\mathbf{P}(A_2|A_1)\ldots\mathbf{P}(A_n|A_1\ldots A_{n-1})$$

if

$$\mathbf{P}(A_1) \neq 0, \ldots, \mathbf{P}(A_1 \ldots A_{n-1}) \neq 0.$$

In the p-adic case situations are possible when it does not follows from the relation $\mathbf{P}(A_1A_2) \neq 0$ that $\mathbf{P}(A_1) \neq 0$ and $\mathbf{P}(A_2) \neq 0$.

Theorem 11.1 (Bayes'). *Suppose the events $A_1, \ldots, A_n, \ldots$ form a complete group of disjoint events,* $\mathbf{P}(A_j) \neq 0$ *and* $\mathbf{P}(B) \neq 0$. *Then* $\mathbf{P}(A_i|B) = \mathbf{P}(A_i)\mathbf{P}(B|A_i)/\sum_j \mathbf{P}(A_j)\mathbf{P}(B|A_j)$.

Let us consider, for instance, a classical *Bayes problem* but applied to the case of a p-adic coin. A coin is a device for which the probability of falling up heads is $x \in \mathbf{Q}_p$ and falling up tails is $(1-x)$. We have a collection of coins and take one of them to carry out an experiment. Then we toss the coin n times. If it comes up heads n_1 times, then what is the probability $q_n(x)$ that the probability of the coin falling up heads is x? We shall study the situation when the distribution of the coins in totality it discrete: we have a countable set of probabilities $\{x_k\}_{k=1}^{\infty}$, $x_j \in \mathbf{Q}_p$, with the probabilities of realizations $\mathbf{P}_0(x_j)$, $\sum_j \mathbf{P}_0(x_j) = 1$. This probability distribution corresponds to the collective M_1 for the p-adic topology of stabilization appearing as a result of the experiment S_1 in the process of which we take one coin of the collection and write down the probability of it falling up heads. Then, for every x_j, we consider a collective $M_2(x_j)$ appearing as a result of the experiments $S_2(x_j)$ consisting in tossing a p-adic coin n times with the probability of it falling up heads x_j. By combining the collectives (see R. von Mises [99],[100]) we form a new sequence of outcomes M corresponding to the experiments S in the process of which we take a coin at random from the collection and toss it n times. In the ordinary frequency theory of probabilities [99],[100] such a sequence is also a collective. In the p-adic case it is, in general, incorrect.

Theorem 11.2. *Suppose that the sequence of probabilities $\{x_k\}$ is bounded in $\mathbb{Q}_p$. Then the sequence M is a p-adic collective.*

Proof. The space of elementary outcomes Ω consists of points $\omega_{km} = \{x = x_k$ and the coin has fallen up heads m times$\}$. As usual, we have $\mathbf{P}(\omega_{km}) = \mathbf{P}_0(x_k)C_n^m x_k^m(1 - x_k^{n-m})$. It remains to show that the series $\sum \mathbf{P}(\omega_{km})$ converges in $\mathbb{Q}_p$ and, to do this, it is sufficient to use the boundedness of $\{C_n^m\}$ in $\mathbb{Q}_p$.

If we use discrete $\mathbb{Q}_p$-valued probability distributions, we can suggest a new approach to the *St. Petersburg paradox* in the probability theory (see, for instance, Borel [16]). We consider a game played by Peter and Paul (the probability of a win in one game is 0.5). The game is played according to the following rules. If Peter wins the first game, then Paul pays him 2 francs and the game is terminated. If Peter loses the first game but wins the second, then Paul pays him 2^2 francs and the game is terminated, ..., if Peter loses $(n-1)$ first games but wins the nth game, then Paul pays him 2^n francs and the game is terminated. The problem consists in determining Peter's stake, i.e., the sum he must pay Paul before starting the game as a compensation for his obligations (a more real variant is the St. Petersburg game for a quit, see Borel [16]). The paradox in the result is that the stake must be infinitely large, namely, $m = \sum_{n=1}^{\infty} \frac{1}{2^n} 2^n = \sum_{n=1}^{\infty} 1$.

The St. Petersburg paradox exerted a considerable effect on the formation of the fundamentals of the probability theory [16]. There are several points of view concerning this paradox. One of the approaches, belonging to Bertrand, is that the infinite answer is considered to be "correct", and, whatever Peter's stake is, he may be sure that in the end he will get rich and will ruin Paul.

In the p-adic case it is possible to realize a somewhat modified version of the St. Petersburg paradox. Suppose that the payments are raised to 4^n. Then in the 2-adic probability theory we can calculate the mean value for Peter's win and get an unexpected answer: $m = \sum_{n=1}^{\infty} \frac{1}{2^n} \cdot 4^n = \sum_{n=1}^{\infty} 2^n = -2$. Naturally, this answer contradicts Bertrand's strategy.

Let us now consider a general model with payments λ^n, $\lambda \in \mathbb{Q}_+$. We introduce a set of a rational numbers $W_p = \{\lambda \in \mathbb{Q}_+ : |\lambda/2|_p < 1\}$, $p = 2, \ldots, \infty$, $|\cdot|_\infty = |\cdot|_{\mathbb{R}}$. If $\lambda \in W_p$, then the series $m(\lambda) = \sum_{n=1}^{\infty} \frac{\lambda^n}{2^n}$ converges in $\mathbb{Q}_p$ defining the mean $m_p : W_p \to \mathbb{Q}$, with $m_p|_{W_p \cap W_q} = m_q|_{W_p \cap W_q}$. By virtue of this condition of compatibility, the mapping $m : W = \cup_{p=2}^{\infty} W_p \to \mathbb{Q}$. If $\lambda \in W_\infty$, then we can apply the ordinary probability theory and if $\lambda \in W_p$, then we can apply the $\mathbb{Q}_p$-valued probability theory.

We can also study a variant of the St. Petersburg paradox with stronger divergencies. Suppose that payments are $\lambda_n = 2^n n\, n!$. Then, for any $p \neq \infty$, the mean is correctly determined, i.e.,

$$m = \sum_{n=1}^{\infty} \frac{2^n n n!}{2^n} = -1.$$

With the use of the p-adic theory we get the possibility of obtaining information from variables which are infinite from the point of view of the real theory.

By analogy we can consider a model with payments λ^n and the probability of Peter's loss $\mu \in \mathbb{Q}$, $0 < \mu < 1$. The mean value of Peter's win $m(\lambda, \mu) = \sum_{n=1}^{\infty} \lambda^n \mu^{n-1}(1 - \mu)$ has sense in the ordinary probability theory if $\lambda\mu < 1$, i.e., it is defined on the set $V_\infty = \{(\lambda, \mu) \in \mathbb{Q}_+^2 : 0 < \mu < 1, \lambda\mu < 1\}$. A problem arises of the extension of the statistical characteristics of the game to a wider set of rational number (we only consider the mean value but can also study, say, the problem of the extension of the dispersion). Let us introduce sets of rational numbers $V_p = \{(\lambda, \mu) \in \mathbb{Q}^2 : \lambda \in \mathbb{Q}_+, \mu \in \mathbb{Q}, |\lambda\mu|_p < 1\}$. If the parameters of the game belong to the set V_p, then the game is described in the framework of the $\mathbb{Q}_p$ probability and thus the mean is extended to the set $V = \cup_{p=2}^{\infty} V_p$.

Chapter VII

Statistical Stabilization with Respect to p-adic and Real Metrics

We present a number of statistical models, where relative frequencies oscillate with respect to the real metric and stabilize with respect to a p-adic one. Thus, we have a possibility to create such random sequences that there is no ordinary frequency probabilities but p-adic probabilities are well defined. Statistical samples with such property we can consider as a new type of randomness. Hence we can consider our investigations as investigations of new types of random sequences. A spectrum of different randomnesses is generated. If we try to classify these random sequences with the aid of Kolmogorov's complexity, then we see that (roughly speaking) the Kolmogorov complexity is increasing as n, where n is a length of a finite segment, for usual random sequences, Mises' kollectives, and it is increasing as $\log_p n$ for p-adic random sequences. Thus, our random sequences are "more simple" then Mises' kollectives. But they are typically not Mises collectives. Hence they are considered as chaotic samples from the point of view of the usual theory of probability.

All these statistical models were realized as statistical experiments on computer. How do we apply these results? Our main idea is to get a new information which was invisible inside of the ordinary probability formalism. For instance, we could distinguish two stochastic objects which are considered as chaotic (not random !) inside of Mises' theory. Relative frequencies may oscillate for both these stochastic objects with respect to the real metric. But in the same time these objects may have different p-adic probabilities. Hence a p-adic probability can be considered as a physical characteristic in the same way as ordinary probability. In Mises' papers was developed the point of view that a probability is the same type of physical characteristics of the object as, for example, mass or electric charge. Consequently, we only introduce a new physical characteristic, a p-adic probability, and nothing more.

1. p-Adic Statistical Simulation

1. statistical stabilization of p-adic digits. In particular case of p-adic topologies of statistical stabilization we can use prime numbers $p = 2, 3, 5, ..., 127, ...$ as parameters of our probability investigations. There are two possibilities to find this parameter for the concrete probability model. First one is to investigate the model by means of theoretical studying and to find p as the consequence of the properties of the investigating model. The second is a statistical simulation with the aid of a computer.

We can understand that it would be impossible to consider an infinite collective in the practical computations. The only possibility is to study finite sequences . And it is also impossible to tell about a limit in p-adic topology in the practical computations (the same situation is also in the ordinary theory of probability over the field of real numbers $\mathbb{R}$). It is only possible to study the stabilization of p-adic digits a_j in the p-adic expansion of the relative frequency $\nu(\omega_i) = n_i/N$. For example, the digit a_0 was stabilized after $N = p^{10}$ observations, a_1 was stabilized after $N = p^{100}$ observations and so on. We said that there is the statistical stabilization of the relative frequency in this case.

If we hope to find the parameter p of the probability model S with the aid of computer we can investigate the statistical stabilization of the relative frequencies in $\mathbb{Q}_2$ then (if there is not the statistical stabilization in $\mathbb{Q}_2$) to consider the statistical stabilization in $\mathbb{Q}_3$, in $\mathbb{Q}_5$, ..., in $\mathbb{Q}_{127}$ and so on.

The concrete statistical simulations have shown to us that if we want to realize our computations for sufficiently large values of p then we would use a sufficiently powerful computer. For example, practically impossible to use the personal computer for the simplest probability models if $p = 127$.

There is no problem to propose a probability model such that there is not the statistical stabilization in $\mathbb{R}$ (so there is not a real valued probability distribution) but there is a good statistical stabilization in $\mathbb{Q}_p$.

2. A toy industrial statistical model. Let us consider some industry (IND) which can be able to produce two kinds of production. For example, IND produce white and red balls. Let us consider such industry process that balls are produced by series of $M = p^k, k = 0, 1, 2...$, balls of the same type. Then let us consider a random process of a production of white and red balls.

The statistical simulation of this process will be organized with the aid of two generators of random numbers (in the sense of the usual theory of probability) :ξ and θ, where $\xi = w$ or r and $\theta = 0$ or 1. Assume there were produced $M = p^k$ balls of some fixed colour in the previous series. Then we use the first generator of random numbers ξ and if $\xi = w$ then IND will produce a series of white balls and if $\xi = r$ then IND will produce a series of red balls ;the length of this new series will be calculated with the aid of the second random generator $\theta : M' = Mp^\theta$ (so it is an extensive process of production). Let the produced balls are mixed in (sufficiently large) box.

We are interested in studying of the global result of IND production. What will be the probability $\mathbf{P}(w)$ to draw a white ball from the box and the probability $\mathbf{P}(r)$ to draw a red ball after a very large period of IND production (in mathematical idealization we can consider an infinite period of IND production) ? To get an answer to this question, we can use the following way. Let $T_1, T_2, ..., T_k, ...$ be the moments of the finishing of the previous series of IND production and the beginning of a new series. Let us compute the relative frequencies $\nu_k(w)$ and $\nu_k(r)$ of white and red balls at the moments $T_k, k = 1, 2, ...$. And let us try to consider limits of $\{\nu_k(w)\}$ and $\{\nu_k(r)\}$ as the answer to this question.

Theorem 1.1. *Let ξ and θ be arbitrary generators of random numbers. Then the limits*

$$\mathbf{P}(w) = \lim_{k\to\infty} \nu_k(w), \mathbf{P}(r) = \lim_{k\to\infty} \nu_k(r)$$

exist in the field of the p-adic numbers.

Proof. Using the equality $|M_{k+1}|_p = |M_k|_p p^{-\theta_{k+1}}$ we get that $|M_k|_p \to 0$, $k \to \infty$ in $\mathbb{Q}_p$. That is why the limits:

$$\mathbf{P}(w) = \lim_{N\to\infty} \sum_{k=1}^{N} \alpha_k M_k / \sum_{k=1}^{N} M_k,$$

$$\mathbf{P}(r) = \lim_{N\to\infty} \sum_{k=1}^{N} (1-\alpha_k) M_k / \sum_{k=1}^{N} M_k,$$

where $\alpha_k = 1$ if $\xi_k = w$ and $\alpha_k = 0$ if $\xi_k = r$, exist in $\mathbb{Q}_p$.

In particular, we get formulas :

$$\mathbf{P}(w) = \sum_{k=1}^{\infty} \alpha_k M_k / \sum_{k=1}^{\infty} M_k,$$

$$\mathbf{P}(r) = \sum_{k=1}^{\infty} (1-\alpha_k) M_k / \sum_{k=1}^{\infty} M_k.$$

Theorem 1.2. *Let θ be an arbitrary generator of random numbers and for a generator ξ probabilities of w and r be not equal to 0. Then the relative frequencies oscillate in the field of real numbers.*

Proof. Let one of the colour exchanges will be at the moment T_j and j be sufficiently large. For example, let us consider the exchange w to r. We have that

$$\nu_j(w) = n_j(w)/N_j, \nu_j(r) = n_j(r)/N_j,$$

where

$$N_j = p^{\theta_1} + \ldots + p^{\lambda_j}, \lambda_j = \sum_{i=1}^{j} \theta_j,$$

and $N_{j+1} = N_J + p^{\lambda_j+\theta_{j+1}}$. Then we get that

$$\nu_{j+1}(w) = n_j(w)/\left(N_j + p^{\lambda_j+\theta_{j+1}}\right) = \nu_j(w)/\gamma,$$

where $\gamma = 1 + u, u = p^{\lambda_j+\theta_{j+1}}/N_j$.. It is obvious that

$$N_j \leq \sum_{i=1}^{\lambda_j} p^i = (p^{\lambda_j+1} - 1)/(p-1)$$

and

$$u \geq p^{\lambda_j+\theta_{j+1}}(p-1)/(p^{\lambda_j+1} - 1).$$

Let us consider the first possibility :$\theta_{j+1} = 1$. Then we get:

$$\gamma \geq 1 + (p-1)/\left(1 - p^{-(\lambda_j+1)}\right) \approx p$$

as j is sufficiently large.

Let us consider the case :$\theta_{j+1} = 2$. Then we get: $\gamma \geq 1 + p(p-1)\left(1 - p^{-(\lambda_j+1)}\right)$ is near $p^2 - p + 1$ as$j$ is sufficiently large.

Of course, to be more careful we must consider the Kolmogorov probability space $(\Omega, F, \mathbf{P}_{Kol})$ and two sequences of usual RV $\xi_n(\omega)$ and $\eta_n(\omega)$. Every sequence consists of independent and equally distributed RV (all notions have the Kolmogorov-probability sense). The real oscillation property of the model does not depend on the probabilities of realization θ_j-values 1 and 2. It depends on non-zero probabilities $q_{Kol}(w)$ and $q_{Kol}(r)$ of ξ_j-values w and r. The relative frequencies are depended on ω : $\nu_j(w) = \nu_j(w;\omega)$ and $\nu_j(r) = \nu_j(r;\omega)$. Using $q_{Kol}(w) \neq 0$, we get that the *Kol*-probability of the event

$$O_w = \{\omega \in \Omega : \exists k = k(\omega) : \xi_i(\omega) = w \forall i > k\}$$

is equal to zero. Using $q_{Kol}(r) \neq 0$, we get that the *Kol*-probability of the same event O_r for the r-realization is also zero. Hence, the *Kol*-probability of the event

$$\{\omega \in \Omega : \forall k \exists j > k : \xi_j(\omega) = w, \xi_{j+1}(\omega) = r\}$$

is equal to 1. Hence, the sequences $\{\nu_j(w;\omega)\}_{j=1}^{\infty}$ and $\{\nu_j(r;\omega)\}_{j=1}^{\infty}$ have not any limit in real topology a.e mod $\mathbf{P}_{Kol}$.

This random IND process we could consider as an algorithm which using two standard (real-sense) random sequences ξ and θ as an initial data generates new random sequence consisting of the colours of balls. But this new sequence is not real-random because its relative frequencies oscillate in $\mathbb{R}$. Thus this algorithm generates new class of random sequences which were not considered in usual theory of probability.

This process of the statistical simulation was realized (with the aid of my student V. Bezgin) on computer and we had seen [67], [92] the stabilization of p-adic digits in the p-adic expansion (of $\nu_j(w)$ and $\nu_j(r)$ and the oscillation of digits in the real expansion. of these frequencies.

There is no problem for everybody to repeat this statistical simulation. The only computer problem is the division of large natural numbers in $\mathbb{Q}_p$.

3. Radiation of multiplets. The previous statistical model can be interpreted in another way. Let us consider a process of radiation instead of the process of IND-production. We have a source $\mathcal{P}$ of radiation of two kinds of particles, R and W. The particles are radiated in multiplets (bursts) consisting of one kind of particles. The kind of particles in a radiated multiplets is a random variable (r.v.) distributed according to a certain law. We suppose that the number of particles in a multiplet is the power of some fixed (prime) number p, i.e., $M = p^k$, $k = 0, 1, \ldots$. We suppose that the radiation process is such that the multiplets grow in size with time according to some stochastic law. The statistical simulation is carried out with the aid of two generators of random numbers: (1) $j = 0, 1$, (2) $i = 1, 2$. If $j = 0$, then a multiplet of R-particles is radiated and if $j = 1$, then a multiplet of W-particles is radiated. The size of a multiplet is determined as follows: the magnitude of the first multiplet is a power p^{l_1} (which can also be random) and if the length of the preceding series was p^{l_m}, then the length of the next series is $p^{l_{m+1}}$, where $l_{m+1} = l_m + i$. We denote by $\nu_m^R = n_m^R/N_m$ the relative frequency of R-particles in the first m bursts and by $\nu_m^W = n_m^W/N_m$ the frequency of W-particles. This relative frequencies have the same properties.

4. A biological model. Suppose that we have a biological organism (BIO) with the following properties: (1) during the production time Δ every BIO generates p BIO and then the production process for him is terminated, (2) the production time Δ is much smaller than the life time T of BIO, i.e. $\Delta << T$. Thus one BIO generates the following biological populations: in time Δ the original BIO generates p BIO and then no longer participates in the production process. By the moment $\tau_1 = \Delta$ there appear $N_1 = 1 + p$ BIO, ..., at the moment $\tau_k = k\Delta$ there are already $N_k = 1 + \ldots + p^k$ BIO. Let us now subject this biological system to a random radiation $\mathcal{P}$. In the generation subjected to the radiation there appear mutants among the offspring. The proportion of these mutants in the offspring of radiated generation is also a random variable. We suppose that the mutants do not take part in the further multiplication and that their life time is congruent with that of a normal individual. A biological population appears which consists of normal individuals and mutants. We are interested in the probability $\mathbf{P}(n)$ that a BIO taken at random from this population is normal, and the probability of a mutant, $\mathbf{P}(m)$. To calculate these probabilities, we calculate the proportions of normal individuals and the mutants in the population after k generations and consider a limit as $k \to \infty$.

Suppose that a random source of radiation is defined by the probabilities $\alpha_0, ..., \alpha_{p-1} \in \mathbb{R}$ (in the sense of the ordinary theory of probability). There α_i is the probability that there are exactly i mutants in the offspring of every BIO of the generation. This number i is the same for all BIO of the generation. Thus the influence of $\mathcal{P}$ at the moment τ_k is the same for every BIO. Sometimes this influence is very small and there is no mutants in the offspring. Sometimes it is very large and there are $p - 1$ mutants among p BIO generated by every BIO. We shall not study the case of the total degeneration, when there will be p mutants in the offspring of every BIO of some generation. To study this case, we need to introduce a probability $\alpha_p \neq 0$.

On the mathematical level this stochastic biological evolution can be described in the following way. Suppose that there is a sequence of independent equally distributed RV $\{\xi_n(\omega)\}$ defined on the ordinary Kolmogorov probability space $(\Omega, F, \mathbf{P}_{Kol})$. These RV assume the values $0, 1, ..., p-1$ with the probabilities $\alpha_0, ..., \alpha_{p-1} \in \mathbb{R}$. If we wish to be more careful, we need to use the symbols $\alpha_0(Kol), ..., \alpha_{p-1}(Kol)$. Thus all our models are the examples of a p-adic stochastic generated with the aid of the ordinary stochastic.

Denote by $G_k(n, \omega)$ a number of normal BIO in the kth generation, $n_k(\omega)$ is a number of all normal BIO to the moment $\tau_k = k\Delta$, symbols $G_k(m, \omega)$ and $m_k(\omega)$ are used for mutants, $N_k(\omega)$ is a volume of the biological population of normal and mutants to the moment τ_k, $\nu_k(n, \omega) = n_k(\omega)/N_k(\omega)$ and $\nu_k(m, \omega) = m_k(\omega)/N_k(\omega)$ are the corresponding frequencies.

Proposition 1.1. *Relative frequencies are defined by*

$$\nu_k(n, \omega) = \sum_{j=0}^{k} G_k(n, \omega)/N_k(\omega) =$$

$$[1 + (p - \xi_1(\omega)) + \cdots + (p - \xi_1(\omega)) \cdots (p - \xi_k(\omega))] \times$$

$$\times [1 + p[1 + (p - \xi_1(\omega)) + \cdots + (p - \xi_1(\omega)) \cdots (p - \xi_{k-1}(\omega))]]^{-1};$$

$$\nu_k(m, \omega) = \sum_{j=0}^{k} G_k(m, \omega)/N_k(\omega) =$$

$$[\xi_1(\omega)) + \cdots + (p - \xi_1(\omega)) \cdots (p - \xi_{k-1}(\omega))\xi_k(\omega)] \times$$

$$\times [1 + p[1 + (p - \xi_1(\omega)) + \cdots + (p - \xi_1(\omega)) \cdots (p - \xi_{k-1}(\omega))]]^{-1}.$$

Theorem 1.3. *Let $\alpha_0 > 0$. Then the sequences of the relative frequencies $\{\nu_k(n, \omega)\}$ and $\{\nu_k(m, \omega)\}$ have the limits in $\mathbb{Q}_p$ a.e.* (mod $\mathbf{P}_{Kol}$).

Proof. Let us consider two attributes A and B. The first one means that there are not any mutations in the generation and the second one means that there are mutations. Thus we can introduce a sequence $\{\eta_k(\omega)\}$ of independent equally distributed RV with the values A and B and corresponding probabilities α_0 and $1 - \alpha_0$. Now

$$O = \{\omega \in \Omega : \{\nu_k(n, \omega)\} \text{has a limit in} \mathbb{Q}_p\} \supset$$

$$\supset \{\omega \in \Omega : \lim_{k\to\infty} |G_k(n,\omega)|_p = 0\},$$

consequently,

$$\Omega \setminus O =\subset \cup_{m=1}^{\infty} \cap_{k=m}^{\infty} D_k, D_k = \{\omega : \eta_k(\omega) = B.\}$$

But

$$\mathbf{P}_{Kol}\Big(\cap_{k+m}^{\infty} D_k\Big) = \lim_{M\to\infty} \prod_{k=m}^{M} (1-\alpha_0).$$

What is about statistical stabilization in the field of real numbers? I have not any mathematical statement in this direction. Together with V. Bezgin I carried out computer calculations connected with this experiment and found out that these frequencies oscillated in the field of real numbers for some values of probabilities α_i and stabilize for another [9]. Here the main problem is a great volume of calculations and it is hard to see a statistical stabilization. This problem was also studied by M. Endo and O. Uoko (see the same paper [9]). They realized a statistical simulation for a long series of generations for some probabilities α_i and there was an oscillation of relative frequencies with respect to the real metric.

We have also calculations with respect to the p-adic metric. There is a statistical stabilization for all α_i with $\alpha_0 \neq 0$. There is no problem to show that relative frequencies are p-adic integers. This is why we shall write their p-adic expansion without negative degree digits beginning directly from zero-degree digit, see the agreement of Ch. 1.

Results of the statistical simulation

(A). $p = 7, \alpha_0 = 0.3(9), \alpha_1 = ...\alpha_5 = 0.0(9)$, $\alpha_6 = 0.1$. We consider 20 generations, present frequencies only for mutants, present only two nonzero digits in the real expansion of frequencies, present all 7-adic digits before the first nonstabilized digit .

Real:

0; 0; 0.49; 0.14; 0.62; 0.59 ;0.21; 0.74; 0.66; 0.28;

0.63; 0.33; 0.52; 0.15; 0.026; 0.0038; 0.00055; 0.61 ;0.67; 0.78.

7-adic :

$$0;\ 0;\ 004366;\ 004361;\ 004400;\ 004014;$$

$$004042;\ 00406326;\ 00406320;$$

$$00406022;\ 00406303;\ 004065403;$$

$$0040654040;\ 00406540412;\ 004065404133;$$

$$004065404134;\ 004065404514;\ 004065404653;\ 004065404062.$$

Thus, the first real digit oscillate, but 9 digits of 7-adic expansion stabilize during 20 generations.

(B). $p = 7, \alpha_0 = 0.2(9), \alpha_1 = 0.1(9), \alpha_2 = ...\alpha_5 = 0.0(9), \alpha_6 = 0.1.$
Real:

$$0.63;\ 0.23;\ 0.042;\ 0.25;\ 016;\ 0.028;\ 013;\ 0.26;\ 0.63;\ 0.33;$$

$$0.52;\ 0.25;\ 0.16;\ 0.38;\ 0.64;\ 0.25;\ 0.51;\ 0.15;\ 0.26;\ 0.17.$$

7-adic:

$$524;\ 560;\ 564;\ 561;\ 5644;\ 5645;\ 5642;\ 5641;\ 5646;\ 5641;\ 5640;$$

$$5641;\ 5640;\ 5643;\ 56422;\ 56424;\ 564210;\ 564216;\ 564215;\ 564216.$$

Here the probability α_0 is less then in (A). and this is why the statistical stabilization in $\mathbb{Q}_7$ is much slowly then in (A). Only 5 digits have stabilized after 20 generations.

(C).$p = 11, \alpha_0 = 0.1(9), \alpha_1 = ... = 0.0(8), \alpha_{10} = 0.$
11-adic:

$$5;\ 4;\ 94;\ 90;\ 95;\ 90;\ 947;\ 94(10);\ 948;\ 949;\ 946;\ 9454;\ 9456;$$

$$9450;\ 9453;\ 9452;\ 9456;\ 9450;\ 94554;\ 94555;\ 9450;\ 9453;\ 945554;$$

$$94550;\ 94554;\ 94559;\ 94558;\ 94553;\ 94555;\ 94558;\ 94557;\ 94555;$$

$$94554;\ 94559;\ 94556;\ 945549;\ 945542;\ 94554(10);\ 945543;\ 945542;$$

$$94554(10)8;\ 94554(10)3;\ 95554(10)58;\ 945554(10)62;\ 94554(10)616;$$

$$94554(10)618;\ 94554(10)616;\ 94554(10)6111;$$

94554(10)6110; 94554(10)6115;

94554(10)6110.

Here only 3 digits of 11-adic expansion have stabilized after 20 steps, 4 digits after 30 steps, 9 after 50 steps.

Now we discuss a possible meaning of these statistical information. How can we use this information in applications ? The first way is the following one. We can distinguish different evolutions of biological populations with the aid of this information. In our model it is the same to distinguish unknown sources of radiation. Let us consider two biological populations. One of them was described by the statistical information of (A) and another was described by (B). If we study the real statistical information of (A) and (B), we see only chaotic oscillations of digits. In particular, we have not any possibility to classify these chaoses. Are these chaoses the same or not ? The first digit oscillates in both cases. Now let us study the p-adic information. There is now problem to see that these two chaoses are different. The first one has the 7-adic probability of mutants equal (approximately) to 00406 and the second one has the 7-adic probability $\mathbf{P}(m) = 56421$. Thus, we have found the strict 7-adic structure among chaotic oscillations. Hence these two biological populations are evolving under the influence of two different sources of radiation.

Intuitively the statistical stabilization in $\mathbb{R}$ is connected with the individual process of mutations, when every BIO can has $0, 1, ..., p-1$ mutants in the offspring with probabilities $\alpha_0, ..., \alpha_{p-1}$ (this stochastic model was also realized on computer by M. Endo and O. Uoko and there was the statistical stabilization in $\mathbb{R}$ and fluctuations in $\mathbb{Q}_p$).

2. On a Definition of the p-adic Frequency Probability

1. General populations of different infinite volumes. The general topological principle of the statistical stabilization of relative frequencies was proposed in the previous Chapter. It was formulated on an intuitive level as a general philosophical conception. Now we wish to formulate this principle on a mathematical level of rigorous. The main problem is the rigorous meaning of the words " statistical stabilization " in the formulation of the general principle. We need to propose the strict sense to a notion "N approaches to infinity." In ordinary Mises' frequency theory (as in applied statistics) to approach to infinity means that N is monotonically increasing with the step $\Delta = 1$. To define a law of a variation of N is the same to define a rule of a calculation of relative frequencies. This rule is standard in ordinary theory of probability, where relative frequencies are calculated after each trial.

But generally speaking, a completion Q_τ of Q with respect to the topology of the statistical stabilization τ has not any order structure. Thus, it would be impossible to define a monotone increasing of N if we consider N as element of Q_τ. Of course, it would be possible to consider relative frequencies as a sequence in topological space $Q_\tau, N \to \nu_N$, where the set of natural numbers is realized as a separate set without any connection to Q_τ (the set of indexes). But if we study more carefully our statistical examples, we understand that this rule is not useful in the p-adic case. We have always calculated relative frequencies not after each trial but after the end of a series of trials. Relative frequencies calculated after each trial do not stabilize both in real and p-adic metrics. Thus, the problem of a rule to calculate relative frequencies must be investigated more carefully.

Some kind of an order structure on Q_p was introduced in Ch. 4 (this structure does not compatible with the algebraic structure of Q_p). But a new problem arises here. There is a great number of different infinite p-adic numbers instead of the unique real infinity. What is an infinite p-adic number to which N approaches ? I think that the unique choice is not natural in the p-adic case. There are different possibilities to "approach to infinity" defined by concrete p-adic probability models. The following probability interpretation of these different infinities may be proposed. An infinite limit of N may be interpreted as the volume of the general population of the considered stochastic model. In the real case there was not any interest to a structure of an infinite general population . It was infinite and that's all. p-adics present to us the possibility to study infinite general populations of different volumes. Thus, if a priori volume $N_\infty \in Q_p$ of the general population of the studying probability model is known, we can consider a sequence $\{N_k\}_{k=1}^\infty$ of natural numbers ($k \to \infty$ in ordinary sense of natural numbers) such that $N_k \to N_\infty$ in Q_p and calculate relative frequencies after $N_1, ..., N_k, ...$ trials. It is the simplest case. But usually there is no possibility to valuate a volume of a general population a priori (see the example of the IND-production). In such case our main idea is to calculate relative frequencies after a series of trials. We shall begin from the case of a priori known volume of the general population.

2. Rules to calculate relative frequencies. Now it would be more convenient to denote the set of all attributes of a random experiment by the symbol $A = \{a_1, ..., a_l, ..., a_s\}$. Let $\mathcal{N} = \{N_k\}_{k=1}^\infty$, $N_k < N_{k+1}$, be a sequence of natural numbers. Here the standard order structure on natural numbers is used. By the definition to fix a statistical rule $\mathcal{N}$ is to calculate relative frequencies after $N_1, N_2, ..., N_l, ...$ trials. Hence if

$$\omega = (\omega_1, \omega_2, ..., \omega_n, ...), \omega_j \in A, \tag{2.1}$$

is a sequence of realizations, then we calculate the relative frequencies $\nu_k(a_l, \omega) = n_k(a_l, \omega)/N_k$ for all $l = 1, ..., s$ and study their limits with respect different topologies on the field of rational numbers .

Here we study only the case when there exists a limit $N_\infty = \lim_{k\to\infty} N_k \neq 0$ in Q_p for one of p. This p-adic number N_∞ can be interpreted as the volume of a general population.

The simplest statistical rule is $\mathcal{N} = \mathcal{N}(p)$, where $N_{k+1} = N_k + p^k, k = 0, 1, ...$ There $N_\infty = -1$. We are not surprised very much to see -1 as the volume of the infinite general population. According to Ch. 4 it is only the symbol to denote an infinite p-adic value.

Now we fix one of the statistical rule $\mathcal{N}$. A sequence ω is said to be a S_p-sequence (or more carefully $S_p\mathcal{N}$-sequence) if limits $\mathbf{P}_p(a_l, \omega) = \lim_{k\to\infty} \nu_k(a_l, \omega)$ exist in Q_p for all $a_l \in A$. Limits $\mathbf{P}_p(a_l, \omega)$ are said to be p-adic probabilities (with respect to the statistical rule $\mathcal{N}$.) The next step must be a definition of p-adic collective to use some kind of random test as in Mises' frequency theory. But at the moment we prefer not to investigate this much more complicated notion and to restrict our considerations to S_p-sequences because we are interested more in p-adics then in randomness. But if you consider our statistical examples you can find a random structure in studied sequences.

Now we can proof the formal frequency propositions of the previous Chapter as the rigorous mathematical statements. Let us consider two fixed labels a_i and a_j and a sequence $\omega = (2.1)$. We shall denote by the symbol $\omega(ij)$ a sequence constructed with the aid of ω by the following rule. The labels a_i and a_j are changed to the single label $u = a_i \vee a_j$ in the sequence ω.

Proposition 2.1. *If a sequence ω is a S_p-sequence, then the sequence $\omega(ij)$ is also a S_p-sequence for every pair of labels a_i and a_j and the additive formula for p-adic probabilities holds :*

$$\mathbf{P}_p(a_i \vee a_j, \omega(ij)) = \mathbf{P}_p(a_i, \omega) + \mathbf{P}_p(a_j, \omega). \tag{2.2}$$

Proof.

$$\mathbf{P}_p(u, \omega(ij)) = \lim_{k\to\infty} n_k(u, \omega(ij))/N_k =$$

$$\lim_{k\to\infty} n_k(a_i, \omega)/N_k + \lim n_k(a_j, \omega)/N_k.$$

Let us consider a set of attributes $B = \{i_1, ..., i_t\}$. It is an event. In the same way we introduce sequence $\omega(i_1, ..., i_t)$ using the label $u = a_{i_1} \vee ... \vee a_{i_t}$. If ω is a S_p-sequence, then $\omega(i_1, ..., i_t)$ is also a S_p-sequence and the additive probability formula holds:

$$\mathbf{P}_p(a_{i_1} \vee ... \vee a_{i_t}, \omega(i_1, ..., i_t)) = \sum_{m=1}^{t} \mathbf{P}_p(a_{i_m}, \omega). \tag{2.3}$$

We shall use also the symbol $\mathbf{P}_p(B, \omega)$ to denote the probability $\mathbf{P}_p(a_{i_1} \vee ... \vee a_{i_t}, \omega(i_1, ..., i_t))$. Thus we get

$$\mathbf{P}_p(C \cup B, \omega) = \mathbf{P}_p(C, \omega) + \mathbf{P}_p(B, \omega), C \cap B = \phi,$$

It is evident that

$$\mathbf{P}_p(A, \omega) = 1 \tag{2.4}$$

Proposition 2.2. *The image of p-adic probability is contained in the ball $\mathcal{U}_R$, $R = |N_\infty|_p^{-1}$ for all S_p-sequences ω.*

In particular, we get that the image of p-adic probability is the ring of p-adic integers for the rule $\mathcal{N}(p)$.

It is evident that $\mathbf{P}_p(a_i, \omega)$ is contained in this ball. We need only to apply the strong triangle inequality to formula (2.3).

Now let us study a case when a priori volume N_∞ of a general population is not known. In this case the following rule to calculate relative frequencies is proposed:

Relative frequencies are calculated when a series of one of the attributes is finished.

Hence we have the following process of the calculation. Let k series of attributes have been finished. We calculate relative frequencies $\nu_k(a_j) = \nu_k(a_j, \omega) = n_k(a_j)/N_k$, where $n_k(a_j)$ is a number of realizations of the attribute a_j during first k series, $N_k = \sum_{j=1}^{s} n_k(a_j)$ is a total number of trials after k series. If the next series is a series of the attribute a_l and the length of this series is $\Delta = \Delta_{l,k+1}$, then $\nu_{k+1}(a_l) = (n_k(a_l) + \Delta)/N_{k+1}$, $\nu_{k+1}(a_j) = n_k(a_j)/N_{k+1}$, $j \neq l$, and $N_{k+1} = N_k + \Delta$. This statistical rule will be denoted by the symbol $\mathcal{NS}$. In the same way we introduce the notion of a S_p-sequence with respect to the statistical rule $\mathcal{NS}$. Now we fix the rule $\mathcal{NS}$ and study the properties of S_p-sequences. At first we shall show that the additive property (2.3) is valid for such sequences.

Proof. Calculate relative frequencies in the sequence $\omega(i_1, ..., i_t)$ after m series . Let there were α series of the attribute $u = a_{i_1} \vee ... \vee a_{i_t}$ and $m - \alpha$ series of other attributes. Let also α series of u were generated by α_1 series of the attribute a_{i_1}, ..., α_t series of the attribute a_{i_t}. Set $\gamma_m = \beta + \alpha_1 + \cdots + \alpha_t$. Then $\gamma_m \to \infty, m \to \infty$, with respect to the ordinary order structure . Hence we have for $j \neq i_l$: $\nu_m(a_j, \omega(i_1, ..., i_t)) = \nu_{\gamma_m}(a_j, \omega) \to \mathbf{P}_p(a_j, \omega)$. Further,

$$\nu_m(u, \omega(i_1, ..., i_t)) = (n_{\gamma_m}(a_{i_1}) + \cdots + n_{\gamma_m}(a_{i_t}))/N_{\gamma_m} =$$

$$\sum_{l=1}^{t} \nu_{\gamma_m}(a_{i_l}, \omega) \to \sum_{l=1}^{t} \mathbf{P}_p(a_{i_l}, \omega).$$

It is also evident that the total probability is equal to 1 and it is the formula (2.4).

Proposition 2.3. *The image of p-adic probability is equal to Q_p for the statistical rule $\mathcal{NS}$.*

Proof. Let $a \in Q_p$. Represent it as $a = \alpha/p^q, \alpha \in Z_p, \alpha = \alpha_0 + \alpha_1 p + \cdots, \alpha_0 \neq 0$. We shall build a S_p-sequence ω with two attributes W and R such that $\nu_k(W, \omega) \to a, k \to \infty$. Introduce the following natural numbers

$$\Delta_k = \alpha_k p^k, \Gamma_k = p^q(p_{k+1} + 1),$$

$$\Delta\Gamma_k = \Gamma_k - \Gamma_{k-1}(\Gamma_{-1} = 0), \delta_k = \Delta\Gamma_k - \Delta_k.$$

Let us consider a sequence of series of W and R. Lengths of series of W are $\Delta_0, ..., \Delta_k, ...$ and lengths of series of R are $\delta_1, ..., \delta_k, ...$ Then every such sequence can be chosen as ω.

We can introduce some kind of random structure if we consider a stochastic rule to choose a series of W or R.

Now let us study frequency conditional probabilities with respect to the statistical rule $\mathcal{NS}$. There we also follow to R.von Mises.

Let $B = \{a_{i_1}, ..., a_{i_t}\}$ be some set of attributes (an event). Consider a sequence (2.1) and construct a new sequence $\omega|B$ using the following rule. We are going along ω and choosing only elements belonging to B. Thus we get

$$\omega|B = (\xi_1, ..., \xi_n, ...), \xi_j \in B, \tag{2.5}$$

Proposition 2.4. *Let ω be a S_p-sequence with respect to $\mathcal{NS}$. Then a sequence $\omega|B$ is also S_p-sequence for each event B with nonzero probability. The following conditional probability formula holds*

$$\mathbf{P}_p(C, \omega|B) = \mathbf{P}_p(C, \omega)/\mathbf{P}_p(B, \omega), \tag{2.6}$$

for each event $C \subset B$.

Proof. Calculate frequencies in (2.5) after k series: $\nu_k(a_{i_l}, \omega|B) = n(a_{i_l}, \omega|B)/N_{k,B}$, where $N_{k,B} = \sum_{q=1}^{t} n_k(a_q, \omega|B)$. Note that the end of kth series in a sequence (2.5) corresponds to the end of $(k+m)$th series in the S_p-sequence (2.1). m series corresponds to attributes $a_j, j \neq$

i_q, with the number of realizations $n_{k+m}(a_j, \omega)$. Further, $n_k(a_{i_q}, \omega|B) = n_{k+m}(a_{i_q}, \omega)$, $q = 1, ..., t$. Thus

$$\nu_k(a_{i_q}, \omega|B) = \Big(n_{k+m}(a_{i_q}, \omega)/N_{k+m}\Big)/\Big(\sum_{q=1}^{n} n_{k+m}(a_{i_q}, \omega)/N_{k+m}\Big),$$

where N_{k+m} is a number of all realizations in (2.1) after $(k+m)$ series. Hence

$$\nu_k(a_{i_q}, \omega|B) = \nu_k(a_{i_q}, \omega)/\nu_k(B, \omega) \to \mathbf{P}_p(a_{i_q}, \omega)/\mathbf{P}_p(B, \omega).$$

Using the symbol $\mathbf{P}_p(C|B, \omega)$ instead of $\mathbf{P}_p(C, \omega|B)$, we get the ordinary conditional probability formula:

$$\mathbf{P}_p(C|B, \omega) = \mathbf{P}_p(C, \omega)/\mathbf{P}_p(B, \omega).$$

3. What Can We Do with p-adic Probabilities?

1. On a calculus of p-adic collectives. We can use p-adic (and more general probabilities) in the same way as in the standard real theory of probability.

The simplest example is the following. Let us consider the IND production with s different kinds of production (s-types of balls). We use the statistical rule $\mathcal{NS}$. Let we have got the probabilities $\mathbf{P}_p(1), ..., \mathbf{P}_p(s) \in \mathbb{Q}_p$ of different types of IND production. More carefully we must use symbols $\mathbf{P}_p(1, \omega), ..., \mathbf{P}_p(s, \omega)$. This means that we have studied a random sequence ω generated by the process of the IND-production. We have seen that it is a S_p-sequence with p-adic probabilities $\mathbf{P}_p(1, \omega), ..., \mathbf{P}_p(s, \omega)$. Now we are interested in the probability of the event that a ball which was drawn from the box had one of the fixed types: $i_1, ..., i_k$. So using formula (2.3), we get the answer

$$\mathbf{P}_p(i_1 \vee ... \vee i_k) = \sum_{t=1}^{k} \mathbf{P}_p(i_t).$$

This trivial example can be generalized to more general case of a topological label space. The probability distribution may be nondiscrete. We use a finite sample in applied investigations . Hence p-adic probabilities $\mathbf{P}_p(1,\omega), ..., \mathbf{P}_p(s,\omega)$ are always rational numbers.

Now let us consider some subgroup $G = (i_1, ..., i_l)$ among labels $(i_1, ..., i_s)$ (for example, the balls of the fixed size but of different colours). Let we have known that a ball b belongs to G. What is the probability of the event $b \in i_t$, where $i_t \in G$? There is no problem with the answer: $\mathbf{P}_p(b \in i_t/b \in G) = \mathbf{P}_p(b \in i_t)/\mathbf{P}_p(G)$.

In the same way we can consider two or more independent IND productions and calculate the distribution of the new collective.

Probability calculus teaches us to compute the probability distributions in derived collectives from given distributions in the collectives from which they have been derived [99], p.14.

In probability calculus the given initial data as well as the final data are probabilities. This initial data belongs to the real number field in the standard theory of probability and it belongs to $U_{\mathbf{Q}_\tau}$ in our general theory.

And we must use the frequency interpretation for our results of theoretical considerations to get applications.

2. New level of R. von Mises' and A.N. Kolmogorov's ideas Main von Mises' idea is the following:

"Probability is a highly mathematical science, but it is not mathematics, just as hydrodynamics is not a branch of the theory of partial differential equations - although it suggests interesting and difficult problems in this field. Our aim in presenting probability theory as a mathematical science is to incorporate into the basic assumptions the idealized basic relations to reality, as this is done in mechanics and in other sciences" ([99], p.44).

Such idealized basic relation to reality is the frequency theory of collectives due to von Mises and it is the theory of probability measures due to Kolmogorov.

There was a very long and a very hard discussion about von Mises' theory and Kolmogorov's theory (see [100] and references in the first chapter of this book). I think that both von Mises' and Kolmogorov's ideas are very and very useful and present investigation has demonstrated new wide field of applications both these ideas.

4. The First Step to the p-adic Theory of Information

1. To find a p-adic structure. Our p-adic theory of probability is the first step to a new p-adic theory of information. What is one of the consequences of our studying of the IND production model by means of the p-adic theory of probability ? It is the possibility to obtain from the initial data an information which would be impossible to obtain from this initial data using the real numbers field calculations. If we try to get the meaningful interpretation of the initial data of the IND model inside of the usual considerations then we get only chaos of oscillations and there is not any information in this initial data from the point of view of the usual theory of information. But there exists some useful information and the problem is only to find this information using the mathematical apparatus adopted to this statistical data. What kind of information can be studied by means of the p-adic theory of information ? As we have seen in the example of the IND model one of the possible applications is the application to the large series (of order p^k) information.

I think that it would be possible to apply p-adic information theory to a large number of problems:

1. Quantum mechanics and field theory (new quantum states). We predict an existence of new quantum states with p-adic probability of realizations. If a quantum system is described by such wave function, then relative frequencies stabilize with respect to the p-adic metric.

2. Industry: processes of an extensive production. Probably, it would be useful in financial statistics. Chaotic oscillations of finances may have a rigorous p-adic structure.

3. Ecology: chaos of oscillations. To study oscillations of different kinds of pollution we could find p-adic structure of this ecological chaos.

4. Biology: biological systems where each biological organism generates p new biological organisms and so on...

2. To apply to cosmic signals ? The p-adic theory of probability is a powerful method of the investigation of the well known data which had considered as a data without any useful information (chaos or noise) after the standard real information investigations.

There are many interesting examples of such information. During last years there were numerous attempts to find an information from another civilizations in cosmic signals. But all these investigations had not any result. But why must we use only real numbers theory of information for these cosmic signals? In the framework of p-adic theory of information it would be possible to investigate cosmic signals again. Probably this attempt would be more successful.

Let us consider a simple example to illustrate our ideas. Somebody is trying to send the communication that some constant $T = -1$. But he is using the 2-adic number field for this communication. He knows that in $\mathbb{Q}_2$ this constant has the next code :

$$T = 1 + 2 + 2^2 + 2^3 + ... + 2^m + ... = 1...1...$$

But anybody who would try to use the real numbers based computer for this cosmic signal would receive infinitely increasing sequence of numbers.

We do not tell that it would be very easy to apply p-adic numbers to the investigation of the cosmic signals. There is a problem of finding the prime number p which is the base of the cosmic signal. But it would be possible as the first attempt to realize computations directly for $p = 2, 3, ...$

5. A Probability Model of a p-adic Coin

We wish to present a mathematical model of a device, which can be considered as a p-adic analogue of usual coin. At the moment there is not a metal-electronic realization of such device. But if you have a financial and technical resources, then there is no problem to construct such metal disk with p-adic probabilities of realizations of its sides. Thus, if a long series trials with this "coin" is realized, then frequencies of realizations of "coin's " sides are oscillate with respect to the real metric and stabilize with respect to the p-adic metric. The parameter p-is included into a construction of a "coin."

1. A description of a p-adic coin. Let we have a metal disk with a label "a" on one side and "b" on another. An internal structure of this disk is not so simple as the structure of the usual metal coin. At first, there is an electric apparatus, which can generate a negative electric charge on the side "a" or on the side "b". A charge can be generated only on one side. So, if it is generated on "a", then it is deleted on "b" and vice versa. Secondly, there is a digital computer (or something similar) with a generator of (pseudo) random numbers, $\xi(\omega) = 0, 1$. For instance, let us consider the case of equal probabilities of realizations of 0 and 1. At the moment we speak about usual generator of random numbers and usual probabilities. Here relative frequencies of realizations of 0 and 1 approach to 1/2 with respect to the real metric. A digital computer organises the work of the electric apparatus according to a following algorithm $\mathcal{A}$.

At first, there is the fixed unit of the time, for instance, $\Delta = 1$ second. If $\xi(\omega) = 0$, then an electric charge (negative) is generated on "a" and it is conserved on "a" during an interval

$$\Delta t' = \Delta t p^2, \tag{5.1}$$

where Δt is a previous interval during which "a" was electrified. And the first interval $\Delta t_1 = \Delta t_1^a = 1$. Thus, we have intervals $\Delta t = 1, p^2, p^4, ..., p^{2k}, ...$ for the side "a". If $\xi(\omega) = 1$, then an electric charge (also negative) is generated on "b" and an interval during which "b" is electrified is also computed with the aid of (5.1), but the first interval $\Delta_1 = \Delta_1^b = p$. Thus, we get intervals $\Delta t = p, p^3, p^5, ..., p^{2k+1}, ...$

This disk with its internal structure is called *a p-adic coin.*

2. A description of coin-tossings. Let us consider a statistical experiment with the p-adic coin. There is an observer $\mathcal{O}$ who does not know anything about an internal structure of this "coin". He see only a metal disk. An observer $\mathcal{O}$ organizes the following experiment with this coin. There is a metal table with the constant positive charge. An observer $\mathcal{N}$ throws the coin on the table. There is one and only one coin-tossing every second. A values of charges are such that the coin always falls in a such way that a top side has not a charge. Besides of all there is yet condition on the algorithms $\mathcal{A}$. This algorithms begins to work after the first coin-tossing.

We are interested in probabilities of realizations of the side "a", $\mathbf{P}(a)$, and the side "b", $\mathbf{P}(b)$.

3. Oscillation of relative frequencies with respect to the real metric. Let $\{\xi_k(\omega)\}$ be a sequence of independent RV, $\xi_k(\omega) = 0, 1$ with probabilities 1/2 (Bernoulli scheme) on the standard probability space $(\Omega, \mathcal{F}, \mathbf{P})$. At the moment we use usual Kolmogorov's definition.

Let us introduce the sums of RV

$$S_n(\omega) = \sum_{k=1}^{n} \xi_k(\omega),$$

$$T_n(\omega) = n - S_n(\omega) = \sum_{k=1}^{n} (1 - \xi_k(\omega)).$$

Set

$$n_m^b(\omega) = \sum_{k=1}^{T_m(\omega)} p^{2(k-1)}$$

$$n_m^a(\omega) = \sum_{k=1}^{S_m(\omega)} p^{2k-1}$$

and $N_m(\omega) = n_m^a(\omega) + n_m^b(\omega)$. Introduce relative frequencies $\nu_m^a(\omega) = n_m^a(\omega)/N_m(\omega)$ and $\nu_m^b(\omega) = n_m^b(\omega)/N_m(\omega)$.

Theorem 5.1. *Relative frequencies $\nu_m^a(\omega)$ and $\nu_m^b(\omega)$ have not a limit in the field of real numbers a.e.*(mod $\mathbf{P}$).

Proof. Let us consider, for example,

$$\nu_m^b(\omega) = n_m^b(\omega)/[n_m^a(\omega) + n_m^b(\omega)] = 1/(1 + \lambda_m(\omega)),$$

where

$$\lambda_m(\omega) = n_m^a(\omega)/n_m^b(\omega) =$$

$$\sum_{k=1}^{S_m(\omega)} p^{2k-1} / \sum_{k=1}^{T_m(\omega)} p^{2(k-1)} =$$

$$p(p^{2S_m(\omega)} - 1)/(p^{2T_m(\omega)} - 1) =$$

$$p(1 - p^{-2S_m(\omega)})/(p^{2(T_m(\omega)-S_m(\omega))} - p^{-2S_m(\omega)}).$$

Using the strong law of large numbers, we get that $p^{-2S_m(\omega)} \to 0, m \to \infty$, a.e.$\omega$. Further, $T_m(\omega) - S_m(\omega) = m - 2S_m(\omega)$, but this expression has not a limit a.e. ω.

We have calculated relative frequencies after every series of 0 or 1. We can also calculate relative frequencies after every trial as in standard probability considerations (for instance, Mises' theory). It is a simple corollary of Theorem 5.1 that limits of these relative frequencies do not exist.

4. Statistical stabilization with respect to the p-adic metric. At first, we consider the statistical rule $\mathcal{NS}$ of the computation of relative frequencies after a series of "a" or "b".

Theorem 5.2. *Relative frequencies $\nu_m^a(\omega)$ and $\nu_m^b(\omega)$ have the limits in Q_p a.e.* (mod $\mathbf{P}$)*.*

Proof. As $S_n(\omega), T_n(\omega) \to \infty$ a.e. (mod $\mathbf{P}$), we get that there exist the limits

$$\mathbf{P}_p(a) = \lim_{m\to\infty} \nu_m^a(\omega) =$$

$$[p/(1-p^2)](1-p) = p/(1+p);$$

$$\mathbf{P}_p(b) = \lim_{m\to\infty} \nu_m^b(\omega) = [1/(1-p^2)](1-p) = 1/1+p.$$

Hence, a.e. $\omega \in \Omega$ are ST_p-sequences with respect to the statistical rule $\mathcal{NS}$. And p-adic frequency probabilities are well defined for all such sequences. These p-adic probabilities do not depend on $\omega \in \Omega$.

For example, if $p = 2$, then 2-adic probabilities of realization of "a" and "b" are equal to 2/3 and 1/3 respectively and if $p = 127$, then $\mathbf{P}_a = 127/128$ and $\mathbf{P}_b = 1/128$.

There is no problem to generalize this p-adic coin to exchange the algorithms $\mathcal{A}$. We can consider another rule for time intervals when "a" or "b" are electrified. For example, $\Delta t_a' = \Delta_a p^4$ and

$\Delta'_b = \Delta_b p^3$. We shall get another p-adic coin with another p-adic probabilities. There is also a possibility to include nonsymmetric generators of (pseudo)random numbers into the digital computer. There is no problem for everybody to propose numerous modifications of such p-adic coin. For instance, you could realize a p-adic analogue of a die with six p-adic probabilities and an oscillation with respect to a real metric.

How can we use p-adic probability in the case of experiments with p-adic coins ? There is the obvious possibility to use p-adics to classify different types of p-adic coins with the aid of statistical experiments. Let an observer $\mathcal{O}$ has two metal disk. These disks are p-adic coins of with different p-adic probabilities $\mathbf{P}_a(1), \mathbf{P}_b(1)$, and $\mathbf{P}_a(2), \mathbf{P}_b(2)$. An observer $\mathcal{O}$ consider the previous statistical experiment with the first and then with the second coin. If he uses ordinary statistics, then he gets a chaotic oscillations of relative frequencies as for the first coin, so for the second one. There is no any possibility to distinguish these coins. Both coins generate only a series of chaotic oscillations. But if $\mathcal{O}$ begins data processing of relative frequencies with the aid of p-adic numbers, he gets immediately different p-adic probabilities. And it is possible to distinguish these coins without any problem.

At these experiment a typical problem p-adic data processing is generated. Before all our data processing we must decide the problem of choosing p. What kind of p-adic numbers are useful in this model ? There is no problem to answer on this question in the case of constructed coin. We see immediately a p^k-series structure of our statistical data. There is a little problem because different series generated by digital computer can be considered as one long series by an observer $\mathcal{O}$.

My student V. Bezgin has realized this p-adic coin model on computer and we have seen an oscillation of digits in decimal expansion of relative frequencies of "a" and "b" during very and very long series of experiments and statistical stabilization of p-adic digits of the same relative frequencies. Of course, a speed of this p-adic stabilization depends very much on different (1/2, 1/2)-generators of pseudo random numbers. But the result is always the same. We had also statistical experiments with nonsymmetric generators. The picture was the same. There was a real digits oscillation and p-adic digits stabilization. But at the moment, there is no any mathematical result on nonsymmetric p-adic coins.

Now let us consider the rule $\mathcal{N}(p)$ of computation of relative frequencies.

Theorem 5.3. *A sequence $\omega \in \Omega$ is a ST_p-sequence with respect to the statistical rule $\mathcal{N}(p)$ a.e*(mod $\mathbf{P}$).

Proof. Denote by m_k^a and m_k^b a number of realizations of "a" and "b" after $M_k = \sum_{j=0}^k p^j = (p^{k+1}-1)/(p-1)$ trails of the p-adic coin and by μ_k^a, μ_k^b are the corresponding frequencies. Set $n^a = \lim_{k\to\infty} n_k^a$ and $n^b = \lim_{k\to\infty} n_k^b$ (see the proof of Theorem 5.2). As $\lim_{k\to\infty} N_k(\omega) = \lim_{k\to\infty} M_k = 1/(1-p)$ a.e. (mod $\mathbf{P}$), then it suffices to show that $\lim_{k\to\infty} m_k^a = n^a$ and $\lim_{k\to\infty} m_k^b = n^b$. We see that

$$M_k = m_k^a + m_k^b = (p + p^3 + \cdots + p^{2U_k-1}) + (1 + p^2 + \cdots + p^{2(V_k-1)}) +$$

+(a part of one of "a"-series of the length p^{2U_k+1} or one of "b"-series of the length p^{2V_k})$= n^a_{h(k,\omega)} + n^b_{g(k,\omega)} + \delta_k$.

Here $U_k = U_k(\omega)$ and $V_k = V_k(\omega)$ and $U_k(\omega), V_k(\omega) \to \infty, k \to \infty$, a.e. (mod $\mathbf{P}$). Consequently, $n^a_{h(k,\omega)} \to n^a$ and $n^b_{g(k,\omega)} \to n^b, k \to \infty$.

Further,

$$\delta_k = (p^{k+1}-1)/(p-1) - \Big[(p^{2V_k}-1)/(p^2-1) + p(p^{2U_k}-1)/(p^2-1)\Big] =$$

$$\Big[(p^{k+1}-1)(p+1) - p^{2V_k} + 1 - p^{2U_k+1} + p\Big]/(p^2-1) =$$

$$(p^{k+2} + p^{k+1} - p^{2V_k} - p^{2U_k+1})/(p^2-1) = p^{l(k)}\Sigma/(p^2-1),$$

where $(p, \Sigma) = 1$.

As $(p^2-1, p) = 1$ and a number δ_k is natural, then (p^2-1) is a divisor of Σ and $\delta_k = p^{l(k)}\gamma_k$, where γ_k is a natural number, $(\gamma_k, p) = 1$. But there are the following possibilities:

a) δ_k is a part of the series of "a" and $m_k^a = n^a_{h(k,\omega)}(\omega) + \delta_k(\omega)$ and $m_k^b(\omega) = n^b_{g(k,\omega)}$;

b) δ_k is a part of the series of "b" and $m_k^b = n_{g(k,\omega)}^b(\omega) + \delta_k(\omega)$ and $m_k^a(\omega) = n_{h(k,\omega)}^a$.

As $|\gamma_k(\omega)|_p \leq 1$ and $l(k,\omega) \to \infty, k \to \infty$, a.e. $(\mathrm{mod}\, \mathbf{P})$, then $\delta_k(\omega) \to 0$ a.e. $(\mathrm{mod}\, \mathbf{P})$. Hence we have obtained the representation :

$$m_k^a = n_{h(k,\omega)}^a(\omega) + z_k(\omega), m_k^b = n_{g(k,\omega)}^b(\omega) + y_k(\omega),$$

where $z_k(\omega), y_k(\omega) \to 0, k \to \infty$,a.e.$(\mathrm{mod}\, \mathbf{P})$. Consequently,

$$\lim_{k\to\infty} m_k^a(\omega) = n^a \quad \text{and} \quad \lim_{k\to\infty} m_k^b(\omega) = n^b \quad \text{a.e.}$$

6. On the Kolmogorov Complexity of p-adic Random Sequences

We have already defined on the mathematical level the notion of the p-adic statistical stabilization. It is the notion of S_p-sequences with respect to the statistical rule $\mathcal{N}$ with the fixed volume N_∞ of the general population or the rule $\mathcal{NS}$, where relative frequencies are calculated after the finish of the series of some attribute. But really it is not a definition of a p-adic collective, because we have not proposed any analog of von Mises' condition of randomness. We cannot apply von Mises' idea to study sequence with the statistical stabilization property which is invariant with respect to subsequences, because the p-adic valuation depends very much on the change of $n(a_l, \omega)$ to few units.

Now we hope to apply to the p-adic case Kolmogorov's idea to define random sequences on the basis of a notion of a complexity of their finite segments. It would be impossible to realize this approach in this book and, this is why, we present only a brief consideration of this subject illustrated on the example of the p-adic coin. We shall also present some philosophical remarks on a nature of stochastics.

1. Kolmogorov complexity. Let Ω be the set of all sequences $\omega = (\omega_j)_{j=1}^\infty, w_j = 0, 1$. We are interested in functions $f : \Omega \to \Omega$. More precisely we are interested in partially recursive functions. You can find main definitions on this subject, for example, in [140]. But we do not wish to use the rigorous logic way and it is sufficient for our purposes to use only the thesis of Church:

A class of algorithmically realized functions (in the intuitive meaning of an algorithm) coincides with a class of all partially recursive functions.

Thus, we are not interested in the mathematical theory of partially recursive functions. It will be sufficient for us to use only algorithms, corresponding to these functions.

As usual, finite vectors $x = (x_1, ..., x_n), x_j = 0, 1$, are called words with respect to the alphabet $\{0, 1\}.l(x) = n$ is the length of the word x, see [140].

Definition 6.1. (A. N. Kolmogorov [95]) *Let $\mathcal{A}$ be an arbitrary algorithm. A complexity of a word x with respect to $\mathcal{A}$ is*

$$K_{\mathcal{A}}(x) = \min l(\pi),$$

where $\{\pi\}$ are the programmes which are able to realize the word x with the aid of $\mathcal{A}$.

This definition depends very much on a structure of $\mathcal{A}$. But A.N. Kolmogorov proved the following theorem, which was a good justification of this definition.

Theorem 6.1. *There exists such algorithm $\mathcal{A}_{\prime}$ (optimal algorithm) that*

$$K_{\mathcal{A}_{\prime}}(x) \preceq K_{\mathcal{A}}(x) \tag{6.1}$$

for every algorithm $\mathcal{A}$.

As usual, (6.1) means that there exists such constant C that

$$K_{\mathcal{A}_{\prime}}(x) \leq K_{\mathcal{A}}(x) + C$$

for all words x. An optimal algorithm $\mathcal{A}_{\prime}$ is not unique.

Definition 6.2. *The complexity $K(x)$ of the word x is equal to the complexity $K_{\mathcal{A}_{\prime}}$ with respect to one fixed (for all considerations) optimal algorithm $\mathcal{A}_{\prime}$.*

A.N. Kolmogorov proposed [95] to use the notion of the complexity of a finite word to try to define a random sequences with the aid of complexities of their finite segments. The idea of Kolmogorov was very natural. He proposed to consider a sequence $\omega \in \Omega$ as a random sequence, if finite segments $(\omega)_n = (\omega_1, ..., \omega_n)$ of this sequence had complexities not much less then n. Thus, a sequence ω is a random sequence in the Kolmogorov sense if it is impossible to find programmes π_n, generating words $(\omega)_n$, with lengths $l(\pi_n) \ll n$. We need a word with a length not less then the length of the segment of ω for coding this segment.

2. An estimate of the complexity of the p-adic coin. We can consider the p-adic coin model in the following way. There is the algorithm $\mathcal{A}$ (now this letter is used for the concrete algorithm considered of the p-adic coin), Turing machine, which transforms every sequence $\omega \in \Omega$ into a new sequence $\xi = f(\omega)$. Let $\{s_1(\xi) < s_2(\xi) < ... < s_n(\xi) < ....\}$ are the moments when 0 is changed to 1 or vice versa. Denote by $m_j = |(\xi)_{s_j}|$ a number of 1 in the word $(\xi)_{s_j}$.

Theorem 6.2. *The following estimate*

$$K((\xi)_{s_j}) \preceq \log_p s_j \tag{6.2}$$

holds a.e. ω *with respect to the standard (real) Bernoulli probability on* Ω.

Proof. On the basis of Theorem 6.1 we have

$$K((\xi)_{s_j}) \leq K_{\mathcal{A}}((\xi))_{s_j}) + C_{\mathcal{A}}.$$

Further,

$$K_{\mathcal{A}}((\xi))_{s_j}) \leq l(f^{-1}((\xi)_{s_j})) = L_j,$$

where $\xi = f(\omega)$. Now $m_j = p(p^{2t+1} - 1)/(p^2 - 1)$, a $s_j - m_j = (p^{2r+1} - 1)/(p^2 - 1)$, where $L_j = t + r$. Thus, $t = (1/2)\log_p \Big[1 + m_j(p^2 - 1)/p\Big] - 1/2$. In the same way we calculate r and get

$$L_j = (1/2)\Big(\log_p \Big[1 + m_j(p^2-1)/p\Big] + \log_p \Big[1 + (s_j - m_j)(p^2-1)\Big]\Big) - 1.$$

Now we use the fact that $m_j \to \infty$ and $s_j - m_j \to \infty, j \to \infty$ a.e. mod Bernoulli probability on Ω (it is ordinary probability measure with real values). Hence

$$L_j \leq (1/2)\Big[\log_p m_j + \log_p(s_j - m_j)\Big] + C$$

As $\sqrt{ab} \leq (a+b)/2$, we get

$$L_j \leq \log_p \sqrt{m_j(s_j - m_j)} + C \leq$$

$$log_p(1/2) + log_p(s_j - m_j + m_j) + C = \log_p s_j + C.$$

Now we wish to present some philosophical considerations about a nature of stochastics. What is the ordinary stochastics ? This is the stochastics which is considered in the standard real theory of probability on the basis of R.von Mises notion of the collective or on the basis of A.N. Kolmogorov notion of the complexity. It is called randomness. On Kolmogorov's language these are sequences of realizations, where a complexity of a finite segment increases as a length of this segment (really, it is not so simple, but we cannot to study this subject more carefully, see, for example, [140]). We have considered one particular example and have seen that sequences with the property of the p-adic statistical stabilization may be much more simple (in the sense of the Kolmogorov complexity) than the ordinary random sequences. Their complexity increases as $\log_p n$ instead of n in the ordinary theory of probability. Consequently, we could try to classify stochastics more carefully and introduce different types of stochastics :n- stochastics (usual real theory of probability), $\log_p n$- stochastics (p-adic theory of probability) and so on. From the standard point of view we have only random sequences, definite sequences (where the Kolmogorov complexity is bounded) and a large middle class sequences, which are not considered as random or definite. The Kolmogorov complexity of these sequences is unbounded, but it is not increasing as a length . We hope that p-adic theory of probability will generate a classification of this middle class stochastics. But now these are only the ideas . Our considerations are based only on the example of the p-adic coin. I think that, as in the ordinary theory of probability, the Kolmogorov complexity is not a good tool to resolve this problem on the mathematical level, see [140]. We need in another types of complexities, see [140]. And, probably, we need to introduce new types of complexities for p-adic investigations.

7. The Statistical Interpretation of Quantum Models with Wave Functions Assuming Values in the Quadratic Extensions of the Field of p-adic Numbers

The frequency probability theory based on the general topological principle of stabilization of relative frequencies makes it possible to suggest a statistical interpretation of the non-Archimedean-valued quantum theories similar to the statistical interpretation of ordinary quantization.

Suppose, for instance, that we are considering the Bargmann–Fock representation $f(z) = \sum_\alpha f_\alpha z^\alpha$, $(f, f) = 1$, $\alpha = (\alpha_1, \ldots, \alpha_n)$.

When the number of trials (energy measurements) tends to infinity, the relative frequency $\nu(\alpha)$ of the realization of energy $E_\alpha = \sum(\alpha_j + \frac{1}{2})$ tends to the probability $|f_\alpha|^2$ in the p-adic topology.

Thus, to my mind the non-Archimedean-valued quantum mechanics corresponds to the consideration of new quantum states with p-adic probabilities of realization.

I can suggest similar statistical interpretation of the non-Archimedean-valued field theory. Suppose, for instance, that we are considering the Bergman–Fock representation $f(z) = \sum_\alpha f_\alpha z^\alpha$, $(f, f) = 1$.

When the number of trials tends to infinity, the relative frequency $\nu(\alpha)$ of α_j particles being in the state z_j tends to the probability $|f_\alpha|^2$ in the p-adic topology.

Thus the non-Archimedean-valued field theory corresponds to the consideration of new quantum-field states with p-adic probabilities of realization.

Note that the assumption on the non-Archimedean structure of the Hilbert space of the quantum theory is not connected with the structure of space–time. We can also consider an ordinary real space–time introducing in this case new quantum states.

For an experimental substantiation of this theory it is necessary to carry out experiments in which the relative frequencies would oscillate in the real topology and would converge in the p-adic topology for one of the fields of p-adic numbers. The prime number p must be either chosen theoretically, on the basis of the properties of the model, or be simulated on a computer by means of sorting out with respect to p. Of course, if p is large, then the latter way is inconvenient.

We can see from the structure of the p-adic statistical models considered in this chapter that the p-adic statistics is suitable for systems with large series whose length tends to zero by the p-adic norm.

Chapter VIII

The p-adic Valued Probability Distributions (Generalized Functions)

Unboundedness of the p-adic Gaussian distribution is the strong reason to create a variant of the p-adic theory of probability, where probabilities belong to spaces of distributions (=generalized functions). This chapter is devoted to this problem. This theory is very similar to the ordinary quantum probability (over the field of real numbers). Both these formalisms develop without measure-theoretical constructions.

1. Axiomatics

Let Ω be a set, $\Phi(\Omega, \mathbf{Q}_p)$ be a topological algebra of functions $\xi : \Omega \to \mathbf{Q}_p$ with the following property:

the composition $f \circ \xi$ belongs to $\Phi(\Omega, \mathbf{Q}_p)$ for all $f \in \mathcal{A}(\mathbf{Q}_p, \mathbf{Q}_p)$ and $\xi \in \Phi(\Omega, \mathbf{Q}_p)$.

Further, let $\mathbf{P} : \Phi(\Omega, \mathbf{Q}_p) \to \mathbf{Q}_p$ be a normalized ($\mathbf{P}(1) = 1$) continuous linear functional with the following condition:

a functional $\mathbf{P}_\xi : \mathcal{A} \to \mathbf{Q}_p$ defined by the equality $\mathbf{P}_\xi(f) = \mathbf{P}(f \circ \xi)$ is continuous for every function $\xi \in \Phi(\Omega, \mathbf{Q}_p)$.

Definition 1.1 *The triple $(\Omega, \Phi(\Omega, \mathbf{Q}_p), \mathbf{P})$ is said to be a generalized probability space of a p-adic valued theory.*

As it was discussed, the field $\mathbf{Q}_p$ plays a role of the segment $[0, 1]$ in a p-adic case. This is why the normalization condition is the only restriction on values of $\mathbf{P}$.

Ω is said to be a sample space of the theory, $\mathbf{P}$ is said to be a probability distribution, functions of the algebra $\Phi(\Omega, \mathbb{Q}_p)$ are called random variables (RV), distribution $\mathbf{P}_\xi$ is said to be a probability distribution of RV ξ.

Of course, our restrictions on the probability distribution $\mathbf{P}$ are very weak. But we do not know until this time what kind of restrictions may be proposed. As we have already discussed, the condition of σ-additivity (on σ-algebra) implied a discreteness of a measure. If only bounded measures are considered as probability distributions, then Gaussian and Volkenborn distributions are excepted from the theory of probability.

There is a number of common points with the quantum theory of probability. In particular, there is not any σ-additive probability measure in the quantum theory of probability.

The simplest example of a probability space is

$$\Omega = \mathbb{Q}_p^n,\ \Phi(\Omega, \mathbb{Q}_p) = \mathcal{A}(\mathbb{Q}_p^n, \mathbb{Q}_p), \mathbf{P} = \mu \in \mathcal{A}'(\mathbb{Q}_p^n, \mathbb{Q}_p)$$

$$\text{with the normalization condition,} \int_{\mathbb{Q}_p^n} 1\ \mu(\mathrm{d}x) = 1.$$

This probability space is convenient to realize a finite number of independent RV. But there is no possibility to build an infinite sequence of independent RV.

The next probability model was proposed in [60] (see section 5 of this chapter). It was a probability space on the basis of the white noise distribution. There was a possibility to build infinite sequences of independent Gaussian RV in this probability space. There we propose a large class of probability spaces. We can build practically all infinite sequences of independent RV on the basis of these probability spaces.

As usual, a mean value of RV $\xi \in \Phi(\Omega, \mathbb{Q}_p)$ is defined as an integral $\mathbf{M}\xi = \int_\Omega \xi(\omega)\mathbf{P}(\mathrm{d}\omega)$ and $\mathbf{M}f(\xi) = \int_{\mathbb{Q}_p} f(x)\mathbf{P}_\xi(\mathrm{d}x)$. Moments are defined by $m_\xi(n) = \mathbf{M}\xi^n$.

Definition 1.2. *A sequence of RV $\{\xi_n\}$ converges with respect to a distribution to RV ξ, $\xi_n \to \xi(D), n \to \infty$, if the sequence of probability distributions $\{\mathbf{P}_{\xi_n}\}$ converges to $\mathbf{P}_\xi$ in the space $\mathcal{A}'(\mathbb{Q}_p, \mathbb{Q}_p)$.*

Definition 1.3 *A characteristic function of RV ξ is defined as the Laplace transform of the distribution $\mathbf{P}_\xi$:*

$$\varphi_\xi(t) = \mathbf{M}e^{t\xi(\omega)} = \int_{\mathbb{Q}_p} e^{tx}\mathbf{P}_\xi(\mathrm{d}x).$$

As $\mathcal{A}' = \cup_r \mathcal{A}'_r$, then for every $\mathbf{P}_\xi \in \mathcal{A}'$ there exists such $r \in \Gamma_{\mathbb{Q}_p}$ that $\mathbf{P}_\xi \in \mathcal{A}'_r$. This is why a characteristic function is well defined for small t.

Remark 1.1. *We omit i in the exponent since in a p-adic case i does not affect the existence of expectation.*

Theorem 1.1. (properties of a characteristic function). *Suppose that ξ is a random variable and $\varphi_\xi(t)$ is its characteristic function. We can name the following properties:*

1. $\varphi_\xi(0) = \int_\Omega \mathrm{P}(\mathrm{d}\omega) = 1$.
2. $\varphi_\xi(t)$ *is a function analytic at zero.*
3. $\mathbf{M}\xi^n = \varphi_\xi^{(n)}(0)$.
4. *A characteristic function uniquely defines the distribution of RV.*
5. *If $f(x) = \int_{\mathbb{Q}_p} \mathrm{e}^{xt}\lambda_f(\mathrm{d}t)$, $\lambda_f \in \mathcal{A}'_0(\mathbb{Q}_p)$, then*

$$\mathbf{M}f(\xi) = \int_{\mathbb{Q}_p} \varphi_\xi(t)\lambda_f(\mathrm{d}t),$$

A random vector is any vector $\xi = (\xi_1, \ldots, \xi_n)$ composed of RV. By analogy with the case of one RV, we introduce moments and distributions of a random vector (a joint distribution of RV). In this case $\mathbf{M}f(\xi) = \int_{\mathbb{Q}_p^n} f(x)\,\mathbf{P}_\xi(\mathrm{d}x)$.

Random variables are independent if their joint distribution has the form $\mathbf{P}_\xi = \mathbf{P}_{\xi_1} \times \ldots \times \mathbf{P}_{\xi_n}$.

The characteristic function of the random vector

$$\varphi_\xi(t) = M\mathrm{e}^{(t,\xi)} = \int_{\mathbb{Q}_p^n} \mathrm{e}^{(t,x)}\mathbf{P}_\xi(\mathrm{d}x)$$

possesses similar properties, in particular,

$$\mathbf{M}\xi_1^{k_1}\cdots\xi_n^{k_n} = \frac{\partial^{k_1+\ldots+k_n}}{\partial t_1^{k_1}\ldots\partial t_n^{k_n}}\varphi_\xi(0).$$

Furthermore, since $\varphi_\xi(0) = 1$, the function $\ln\varphi_\xi(t) \in \mathcal{A}_0(\mathbb{Q}_p^n)$ and we can introduce cumulants of the random vector ξ which are connected with the mixed moments by standard formulas (these formulas are purely algebraic in character and do not depend on the choice of a number field).

Theorem 1.2. *For the components of the random vector $\xi = (\xi_1, \ldots, \xi_n)$ to be independent, it is necessary and sufficient that its characteristic function be the product of the characteristic functions of the components.*

Proof. We shall demonstrate how the method of characteristic functions can be used in this theory. Let us prove the sufficiency. If the function $f(x) = f_1(x_1)\ldots f_n(x_n)$, then, in accordance with property 5 of a characteristic function (for a random vector), we have

$$\mathbf{M}f(\xi) = \int\limits_{\mathbb{Q}_p^n}\varphi_\xi(t)\lambda_f(dt) = \int\limits_{\mathbb{Q}_p^n}\varphi_\xi(t)\lambda_{f_1}(dt_1)\ldots\lambda_{f_n}(dt_n) =$$

$$= \int\limits_{\mathbb{Q}_p}\varphi_{\xi_1}(t_1)\lambda_{f_1}(dt_1)\ldots\int\limits_{\mathbb{Q}_p}\varphi_{\xi_n}(t_n)\lambda_{f_n}(dt_n) = \mathbf{M}f_1(\xi_1)\ldots\mathbf{M}f_n(\xi_n).$$

We can also use a characteristic function to prove the following proposition.

Proposition 1.1. *The distribution of the sum of the independent random variables $\xi = \xi_1 + \ldots + \xi_n$ is equal to the convolution of the distributions of the terms $\mathbf{P}_\xi = \mathbf{P}_{\xi_1} * \ldots * \mathbf{P}_{\xi_n}$.*

The random vector ξ is Gaussian if its characteristic function is a quadratic exponent

$$\varphi_\xi(t) = \exp\left\{\frac{1}{2}(Bt, t) + (m, t)\right\},$$

where $m \in \mathbf{Q}_p^n$ and B is a symmetric matrix. In other words, the distribution of this random vector is a Gaussian distribution on $\mathbf{Q}_p^n$, as in the standard theory of probability, we use the symbol $N(m, B)$ to denote the class of Gaussian distributions with parameters m and B.

It immediately follows from the definition that for the Gaussian random vector $\xi = (\xi_1, \ldots, \xi_n)$

$$\mathbf{M}\xi(\omega) = m, \qquad B_{kl} = \mathrm{cov}\,(\xi_k, \xi_l).$$

Theorem 1.3. *The random vector $\xi = (\xi_1, \ldots, \xi_n)$ is Gaussian if and only if the random variable (ξ, λ) is Gaussian for any vector λ.*

To prove this theorem, it is sufficient to consider a characteristic function.

Theorem 1.4. *The non-correlatedness of the components of a Gaussian vector is equivalent to their independence.*

We must again consider a characteristic function.

We see that many important properties of Gaussian random variables are retained in the p-adic case. However, nothing is known about many other properties of p-adic Gaussian random variables.

The following are open questions:

1. *Is the central limit theorem valid in the p-adic case?*
2. *Do we have a p-adic variant of the Levi–Cramer theorem?*

Let ξ, η be independent random variables. If their sum $\xi + \eta$ is a Gaussian random variable, then are ξ and η Gaussian?

3. *Are the characteristics of Gaussian random variables valid in the p-adic case which are based on the properties of linear combinations?* (See, for instance, Hida's monograph [35].)

Definition 1.4. (An analogue of the uniform distribution) *RV ξ is said to be Volkenborn RV if $\mathbf{P}_\xi$ is the Volkenborn distribution.*

2. Probability Distributions on Spaces of p-adic Sequences

We shall consider probability spaces constructed as infinite products of one-dimensional probability distributions. The sample space will be $\mathbb{Q}_p^\infty = \{x = (x_1, ..., x_n, ...) : x_j \in \mathbb{Q}_p\}$.

Let Λ be the set of all multi-indexes $\alpha = (\alpha_1, ..., \alpha_n, ...)$ with $\alpha_j = 0$ except for a finite number of j. Set $i(\alpha) = \max\{m : \alpha_m \neq 0\}$ for $\alpha \in \Lambda$. Further, series $\sum_{\alpha\in\Lambda} a_\alpha$ are considered as

$$\sum_{n=0}^{\infty} \sum_{|\alpha|=n; i(\alpha)=1}^{\infty} a_\alpha.$$

Let us denote by $\mathcal{P}_c \equiv \mathcal{P}(\mathbb{Q}_p^\infty, \mathbb{Q}_p)$ the space of cylindrical polynomials $q : \mathbb{Q}_p^\infty \to \mathbb{Q}_p$,

$$q(x) = \sum_{|\alpha|\leq N} \sum_{i(\alpha)\leq M} q_\alpha x^\alpha, q_\alpha \in \mathbb{Q}_p.$$

Definition 2.1. *Linear functionals $A : \mathcal{P}_c \to \mathbb{Q}_p$ are said to be cylindrical distributions $\mathbb{Q}_p^\infty$.*

Theorem 2.1. *Let $\{\mu_n\}$ be a sequence of normalized distributions of the class $\mathcal{A}'$. Then the equality*

$$\int_{\mathbb{Q}_p^\infty} q(x)\mu(\mathrm{d}x) = \sum q_\alpha \int_{\mathbb{Q}_p^{i(\alpha)}} x^\alpha \mu_1 \otimes \cdots \otimes \mu_{i(\alpha)}(\mathrm{d}x)$$

defines a cylindrical distribution.

This cylindrical distribution is denoted by $\prod \mu_j$. Of course, the class of functions of an infinite argument $\mathcal{P}_c$ is very restricted. It is a problem to continue cylindrical distributions to more extended spaces. Here we need to use a structure of distributions μ_n in a direct product.

Let us consider a space of formal power series $\mathcal{A}_f \equiv \mathcal{A}_f(\mathbb{Q}_p^\infty, \mathbb{Q}_p)$:

$$\sum_{n=0}^{\infty} \sum_{|\alpha|=n; i(\alpha)=1}^{\infty} f_\alpha x^\alpha = \sum_{n=o}^{\infty} f_{(n)}(x). \tag{2.1}$$

The problem is to continue a cylindrical distribution μ from $\mathcal{P}$ to a subspace of $\mathcal{A}_f$ with the aid of the equality:

$$\int_{\mathbb{Q}_p^\infty} f(x)\mu(\mathrm{d}x) = \sum_{\alpha\in\Lambda} f_\alpha \int_{\mathbb{Q}_p^{i(\alpha)}} x^\alpha \mu_1 \otimes \cdots \otimes \mu_{i(\alpha)}(\mathrm{d}x) \tag{2.2}$$

Let us use the symbol $\mathcal{U}_R^\infty, R = (R_1, ..., R_n, ...), R_j \in \Gamma_{\mathbf{Q}_p}$ to denote a "ball" $\{x \in \mathbf{Q}_p^\infty : |x_j|_p \leq R_j\} = \mathcal{U}_{R_1} \times \cdots \times \mathcal{U}_{R_n} \times \cdots$. Of course, there is no any topology generated with the aid of these balls.

Let $\mathcal{A}(\mathcal{U}_R^\infty, \mathbf{Q}_p)$ be the subspace of $\mathcal{A}_f(\mathbf{Q}_p^\infty, \mathbf{Q}_p)$ consisting of such functions $f(x)$ that power series (2.1) converges in the following sense:

$$\lim_{n\to\infty} \sup_{x\in\mathcal{U}_R^\infty} |f_{(n)}(x)|_p = 0.$$

Such functions are said to be analytic on the "ball" $\mathcal{U}_R^\infty$.

Theorem 2.2. *The space $\mathcal{A}(\mathcal{U}_R^\infty, \mathbf{Q}_p)$ coincides with a subspace of $\mathcal{A}_f(\mathbf{Q}_p^\infty, \mathbf{Q}_p)$ consisting of all functions f satisfying the conditions:*

1. $\lim_{|\alpha|=n; i(\alpha)\to\infty} |f_\alpha|_p R^\alpha = 0$ *for every natural n;*

2.$\lim_{\to\infty} \max_{|\alpha|=n; i(\alpha)} |f_\alpha|_p R^\alpha = 0$.

Proof. A. Let $f \in \mathcal{A}_f$ satisfies to conditions 1 and 2. We have $\lim_{|\alpha|=n;i(\alpha)\to\infty} |f_\alpha x^\alpha|_p = 0$ for every $x \in \mathbf{Q}_p^\infty$. Thus, the series

$$f_{(n)}(x) = \sum_{|\alpha|=n; i(\alpha)=1}^{\infty} f_\alpha x^\alpha$$

converges on $\mathcal{U}_R^\infty$. Further,

$$\sup_{x\in\mathcal{U}_R^\infty} |f_{(n)}(x)|_p \leq \sup_{x\in\mathcal{U}_R^\infty} \max_{|\alpha|=n; i(\alpha)} |f_\alpha x^\alpha|_p \to 0, n \to \infty.$$

B. Let now $f \in \mathcal{A}(\mathcal{U}_R^\infty, \mathbf{Q}_p)$. At first, the series of $f_{(n)}(x)$ converges for every $x \in \mathcal{U}_R^\infty$. As $R_j \in \Gamma_{\mathbf{Q}_p}$, then there exist $a_{R_j} \in \mathbf{Q}_p$ such that $|a_{R_j}|_p = R_j$. Let $x_R = (a_{R_j})_{j=1}^\infty$; then $x_R \in \mathcal{U}_R^\infty$ and $\lim_{|\alpha|=n;i(\alpha)\to\infty} |f_\alpha|_p R^\alpha = 0$. Thus, we have proved 1. Using this, we get, in particular, that

$$\max_{|\alpha|=n; i(\alpha)} |f_\alpha|_p R^\alpha = |f_{\beta(n)}|_p R^{\beta(n)} < \infty$$

for every n.

Introduce finite vectors $x(n) = x(R, \beta(n)) : x(n)_j = 0$ if $\beta(n)_j = 0$ and $x(n)_j = a_{R_j}$ if $\beta(n)_j \neq 0$. Then

$$|f_{\beta(n)}|_p R^{\beta(n)} = |f_{(n)}(x(n))|_p \leq \sup_{x\in\mathcal{U}_R^\infty} |f_{(n)}(x(n))|_p \to 0, n \to \infty.$$

But it is the second condition.

Let us introduce on $\mathcal{A}(\mathcal{U}_R^\infty, \mathbb{Q}_p)$ the non-Archimedean norm $||f||_R = \max_{\alpha\in\Lambda} |f_\alpha|_p R^\alpha$.

Proposition 2.1. *The space $\mathcal{A}(\mathcal{U}_R^\infty, \mathbb{Q}_p)$ is a non-Archimedean Banach algebra.*

A proof is based on direct computations with power series.

Theorem 2.3. *Let $\{\mu_n\}$ be a sequence of normalized distributions of the class $\mathcal{A}'$ and $\mu_n \in \mathcal{A}'_{R_n}$ and $\lambda = \sup_n ||\mu_n||_{R_n} \leq 1$. Then a cylindrical distribution $\mu = \prod \mu_n$ is continued to a bounded linear functional on $\mathcal{A}(\mathcal{U}_R^\infty, \mathbb{Q}_p), R = (R_1, R_2, ...)$.*

Proof. We show that (2.2) defines a bounded linear functional on $\mathcal{A}(\mathcal{U}_R^\infty, \mathbb{Q}_p)$. At first, consider series :

$$I_{(k)} = \sum_{|\alpha|=k;i(\alpha)=1}^{\infty} f_\alpha \int_{\mathbb{Q}_p^{i(\alpha)}} x^\alpha \mu(\mathrm{d}x).$$

We get

$$|f_\alpha \int_{\mathbb{Q}_p} x_1^{\alpha_1} \mu_1(\mathrm{d}x_1)... \int_{\mathbb{Q}_p} x_{i(\alpha)}^{\alpha_{i(\alpha)}} \mu_{i(\alpha)}(\mathrm{d}x_{i(\alpha)})|_p \leq |f_\alpha|_p R^\alpha \to 0, i(\alpha) \to \infty.$$

Hence all these series converge and $|I_{(k)}|_p \leq \max_{|\alpha|=k} |f_\alpha|_p R^\alpha$. Consequently, series (2.2) converges and we get the estimate:

$$|\int f(x)\mu(\mathrm{d}x)|_p \leq ||f||_R.$$

Theorem 2.4. *Let a function f be in $\mathcal{A}(\mathbb{Q}_p, \mathbb{Q}_p)$ and a function g be in $\mathcal{A}(\mathcal{U}_R^\infty, \mathbb{Q}_p)$. Then the composition $\phi = f \circ g$ belongs to $\mathcal{A}(\mathcal{U}_R^\infty, \mathbb{Q}_p)$.*

Proof. The proof is a long series of manipulations with power series. Formally,

$$\phi(x) = \sum_{m=0}^{\infty} \sum_{|\gamma|=m;i_\gamma=1}^{\infty} \Big[\sum_n f_n \sum_{\alpha^1+\cdots+\alpha^n=\gamma} g_{\alpha^1} \cdots g_{\alpha^n}\Big] x^\gamma. \qquad (2.3)$$

At first, we show that coefficients ϕ_γ of x^γ are well defined:

$$|\phi_\gamma|_p \leq |f_n|_p ||g||_R^{-\gamma} \to 0, n \to \infty.$$

Further, we show that functions $\phi_{(n)}(x)$ are well defined:

$$|\phi_\gamma|_p R^\gamma \le \max_n |f_n|_p \max_{\alpha^1+\cdots+\alpha^n=\gamma} \prod_{j=1}^n |g_{\alpha^j}|_p R^{\alpha_j}.$$

But there exists α^j with an index equals to an index of γ. Let us choose such N_ϵ that if $i_\gamma \ge N_\epsilon$, then $|g_\gamma|_p R^\gamma < \epsilon$ for $\gamma = m$. Then $|g_{\alpha_j}|_p R^{\alpha_j} < \epsilon$ and $|\phi_\gamma|_p R^\gamma F2 \le \epsilon ||f||_{||g||_R} ||g||_R^{-1}$.

To finish the proof, we need to show the convergence of (2.3). First proposition in this direction is the following: $\forall \epsilon > 0 \exists M_\epsilon$: $\forall i_\gamma \ge M_\epsilon$: $|g_\gamma|_p R^\gamma < \epsilon$ for all m. Indeed, $\forall \epsilon > 0 \exists m_\epsilon : \forall m \ge m_\epsilon$: $\max_{|\gamma|=m} |g_\gamma|_p R^\gamma \le \epsilon$, and for every $m, 0 \le m < m_\epsilon$ there exists $N_{m\epsilon} : \forall i_\gamma \ge N_{m\epsilon}, |\gamma| = m, |g_\gamma|_p R^\gamma < \epsilon$. To finish these considerations, we choose $M_\epsilon = \max_m N_{m\epsilon}$.

Now we need to estimate the expression:

$$\Delta_m = \max_{|\gamma|=m; i_\gamma<\infty} \max_n |f_n|_p \max_{\alpha^1+\cdots+\alpha^n=\gamma} \prod_{j=1}^n |g_{\alpha^j}|_p R^{\alpha_j} =$$

$$\max[\Delta_m(M_\epsilon), \Delta'_m(M_\epsilon)],$$

where

$$\Delta_m(M_\epsilon) = \max_{|\gamma=m|, i_\gamma \le M_\epsilon} [\ldots]; \Delta'_m(M_\epsilon) = \max_{|\gamma=m|, i_\gamma > M_\epsilon} [\ldots].$$

If $i_\gamma > M_\epsilon$, then there exists such $j, 1 \le j \le n$, that $i_{\alpha^j} > M_\epsilon$. Consequently,

$$\Delta'_m(M_\epsilon) \le \max_n |f_n|_p ||g||_R^{n-1} \le \epsilon ||f||_{||g||_R} ||g||_R^{-1} \to, \epsilon 0.$$

As $f \in \mathcal{A}(\mathbb{Q}_p, \mathbb{Q}_p)$, then $\forall \epsilon > 0 \exists n_\epsilon \forall n \ge n_\epsilon : |f_n|_p ||g||_R^n < \epsilon$. Hence

$$\Delta_m(M_\epsilon) \le \max_{|\gamma|=k} \max(\epsilon, \max_{n \le n_\epsilon} |f_n|_p \max_{\alpha^1+\cdots+\alpha^n=\gamma, i_\gamma \le M_\epsilon} \prod_{j=1}^n |g_{\alpha^j}|_p R^{\alpha_j}).$$

In the same way we can prove that for small ϵ

$$\Delta_m(M_\epsilon) \le \epsilon \max(1, ||f||_{||g||_R} ||g||_R^{-1}).$$

Theorem 2.5. *Let a functional ν belongs to $\mathcal{A}'(\mathcal{U}_R^\infty, \mathbb{Q}_p)$ and a function g belongs to $\mathcal{A}(\mathcal{U}_R^\infty, \mathbb{Q}_p)$. Then the functional $\nu_g : \mathcal{A}(\mathbb{Q}_p, \mathbb{Q}_p) \to \mathbb{Q}_p$, $\nu_g(f) = \nu(f \circ g)$, belongs to the space $\mathcal{A}'(\mathbb{Q}_p, \mathbb{Q}_p)$.*

Proof. Using the previous Theorem, we get that the functional ν_g is defined on the space $\mathcal{A}(\mathbb{Q}_p, \mathbb{Q}_p)$. Further,

$$|\nu_g(f)|_p \leq \max_n |f_n|_p |\nu(g^n)|_p \leq ||\nu|| \max_n |f_n|_p ||g^n||_R \leq$$

$$||\nu|| \max_n |f_n|_p ||g||_R^n.$$

Let $\mathbf{P}$ be a normalized bounded linear functional on the functional space $\mathcal{A}(\mathcal{U}_R^\infty, \mathbb{Q}_p)$. Then the triple $(\mathcal{U}_R^\infty, \mathcal{A}(\mathcal{U}_R^\infty, \mathbb{Q}_p), \mathbf{P})$ is a generalized probability space of the p-adic valued probability theory. In particular, it is valid for distributions of the type $\mathbf{P} = \prod \mathbf{P}_j$, where $\mathbf{P}_j$ are probability distributions on $\mathbb{Q}_p$ satisfying the conditions of Theorem 2.3.

Example 2.1. (A probability space of Volkenborn RV). As the Bernoulli numbers B_n satisfy the estimate $|B_n|_p \leq p$, we get the estimate $|\int x^n \mathrm{d}x|_p \leq p$ and $|(\mathrm{d}x, f)|_p \leq ||f||_p$ for $f \in \mathcal{A}$. Thus, $||\mathrm{d}x||_p \leq 1$. If we choose $\mu_n = \mathrm{d}x$ and $R_n = p$ for all n, then the conditions of Theorem 2.3 are satisfied. Hence $(\mathcal{U}_p^\infty, \mathcal{A}(\mathcal{U}_R^\infty, \mathbb{Q}_p), \mathbf{P} = \prod_{i=1}^\infty \mathrm{d}x)$ is a probability space. Set $\xi_n(\omega) = \omega_n$ for $\omega = (\omega_1, ..., \omega_n, ...) \in \mathcal{U}_R^\infty$. It is an infinite sequence of independent Volkenborn RV.

Example 2.2 (A probability space of Gaussian RV). Let $\{\gamma_{0,b_n}\}_{n=1}^\infty$ be a sequence of Gaussian distributions with zero mean values and covariations b_n. Using the estimate of Gaussian moments:

$$|\int x^{2n} \gamma_{0,b}(\mathrm{d}x)|_p = |b^n (2n-1)!!|_p \leq |b|_p^n,$$

we get that the conditions of Theorem 3 are satisfied for $R_n \geq |b_n|_p^{1/2}$ and $\mu_n = \gamma_{b_n}$. Thus, $(\mathcal{U}_p^\infty, \mathcal{A}(\mathcal{U}_R^\infty, \mathbb{Q}_p), \mathbf{P} = \prod_{i=1}^\infty \gamma_{b_n})$ is a probability space. Set $\xi_n(\omega) = \omega_n$. It is an infinite sequence of independent Gaussian RV with laws $N(0, b_n)$.

3. A Limit Theorem

Here we need in some additional results on analytic functions in the p-adic case. Let us consider analytic functions on $\mathbb{C}_p$. We shall use balls in $\mathbb{C}_p$ and $\mathbb{Q}_p$ in the same time. This is why we must use different symbols to denote these balls. We shall use the symbol $\mathcal{W}_R$ to denote a ball in $\mathbb{C}_p$ and the symbol $\mathcal{U}_R$ to denote a ball in $\mathbb{Q}_p$. A radius R belongs to $\Gamma_{\mathbb{C}_p} = \{g = p^t : t \in Q\}$ in the first case and R belongs to $\Gamma_{\mathbb{Q}_p} = \{g = p^t : t = 0, \pm 1, \pm 2, ..., \}$ in the second case.

According to the definition a function $f : \mathcal{W}_R \to \mathbb{C}_p$ is analytic iff $\lim_{n\to\infty} |f|_p R^n = 0$ (we use the same symbol $|\cdot|_p$ for valuations on $\mathbb{Q}_p$ and $\mathbb{C}_p$.) The space $\mathcal{A}(\mathcal{W}_R, \mathbb{C}_p)$ of analytic functions is endowed with the norm $||f||_R = \max_n |f_n|_p R^n$ (see Chapter 2). Let us introduce another norm on this space

$$|||f|||_R = \sup_{z\in\mathcal{W}_R} |f(z)|_p.$$

It is also a non-Archimedean norm on $\mathcal{A}(\mathcal{W}_R, \mathbb{C}_p)$.

We shall use the following theorem (see, for example, [106]):

The equality $||f||_R = |||f|||_R$ *holds for all* $f \in \mathcal{A}(\mathcal{W}_R, \mathbb{C}_p)$ *and* $R \in \Gamma_{C_p}$.

The main fact which is used to prove this theorem is that the field $\mathbb{C}_p$ is not locally compact. For example, this theorem does not valid in $\mathbb{Q}_p$.

An arbitrary function $f \in \mathcal{A}(\mathcal{U}_R, \mathbb{Q}_p)$ may be extended to a function $f \in \mathcal{A}(\mathcal{W}_R, C_p)$(we shall use the same symbol for these functions). Consequently, we get $||f||_R = \sup_{z\in\mathcal{W}_R} |f(z)|_p, R \in \Gamma_{\mathbb{Q}_p}$, for every function $f \in \mathcal{A}(\mathcal{U}_R, \mathbb{Q}_p)$.

Theorem 3.1. *Let* $\{\xi_{nk}(\omega)\}_{k=1}^n, n = 1, 2, ...,$ *be a sequence of independent RV satisfying to the condition :*

There exists sequence of real numbers $\{a_{nk}\}_{k=1}^n, n = 1, 2, ...,$ *which approaches zero* , $n \to \infty$, *uniformly with respect to* k. *And the estimate:*

$$|m_{\xi_{nk}}(j)|_p \le (a_{nk})^j, j = 1, 2, ..., \tag{3.1}$$

holds.

Then the sequence of RV $S_n = \sum_{k=1}^n \xi_{nk}$ converges with respect to a distribution to RV $\xi(\omega)$ with the probability distribution δ, where δ is the Dirac measure concentrated in $\{0\}$.

Proof. The distribution of RV S_n is equal to the convolution $\mathbf{P}_{S_n} = \mathbf{P}_{\xi_{n1}} * \cdots * \mathbf{P}_{\xi_{nn}}$. Consider characteristic functions

$$\varphi_{\xi_{nk}}(t) = \sum_{j=0}^{\infty} \frac{t^j}{j!} M_{\xi_{nk}}(j).$$

Let us fix $R \in \Gamma_{\mathbb{Q}_p}$. Using the estimate (1), we get that

$$\lim_{j\to\infty} R^j |m_{\xi_{nk}}(j)/j!|_p = 0$$

beginning from some n. Thus, all functions $\varphi_{\xi_{nk}}(t)$ belong to $\mathcal{A}(\mathcal{U}_R, \mathbb{Q}_p)$ beginning from some n. Hence the characteristic function $\varphi_{S_n}(t) = \prod_{k=1}^n \varphi_{\xi_{nk}}(t)$ belongs to $\mathcal{A}(\mathcal{U}_R, \mathbb{Q}_p)$. According to the remark before this theorem, we can continue characteristic functions $\varphi_{\xi_{nk}}$ and φ_{S_n} to analytic functions on the ball $\mathcal{W}_R$.

Further, $\varphi_{\xi_{nk}}(z) = 1 + A_{nk}(z)$. We show that $A_{nk} \to 0, n \to \infty$, uniformly with respect to k in the space $\mathcal{A}(\mathcal{W}_R, \mathbb{C}_p)$:

$$|||A_{nk}|||_R = \sup_{z\in\mathcal{W}_R} |A_{nk}(z)|_p \le$$

$$\sup_{z\in\mathcal{W}_R} \max_{1\le j<\infty} |z|_p^j p^{j/(p-1)} |m_{\xi_{nk}}(j)|_p \le (R\, p^{1/(p-1)}\, a_{nk})^j.$$

On the basis of (3.1) we get that $\forall \epsilon > 0 \exists N_\epsilon : \forall n \ge N_\epsilon\ a_{nk} \le \epsilon/Rp^{1/(p-1)}$. Consequently, $|||A_{nk}|||_R \le \epsilon$.

Further, $\varphi_{S_n}(z) = 1 + L_n(z)$. And we have:

$$|||L|||_R = |||\sum_{k=1}^n A_{nk} + \sum_{k\ne l} A_{nk}A_{nl} + ...|||_R \le$$

$$\max(\max_{1\le k\le n} |||A_{nk}|||_R, \max_{1\le k\ne l\le n} |||A_{nk}|||_R\, |||A_{nl}|||_R, ...) \le \epsilon.$$

if $n \ge N_\epsilon$. Thus, $|||L_n|||_R \to 0, n \to \infty$. And, in particular, $L_n \to 0$ in the space $|\mathcal{A}(\mathcal{U}_R, \mathbb{Q}_p)$. To finish the proof we use Theorem 3.1.

Theorem 3.2 *Let $\{\xi_n(\omega)\}_{n=1}^{\infty}$ be a sequence of independent equally distributed RV with zero mean values and $\{c_n\}_{n=1}^{\infty}$ be a sequence of p-adic numbers and $|c_n|_p \to \infty$. Then RV $T_n = S_n/c_n \to \xi(D)$, where $\mathbf{P}_\xi = \delta$.*

Proof. As $\mathbf{P}_{\xi_k} \in \mathcal{A}'$, then $|m_{\xi_k}(j)|_p \leq c^j$. Consequently,

$$|m_{(\xi_k/c_n)}(j)|_p \leq (c/|c|_p)^j \to 0, n \to \infty.$$

In particular, $\frac{\xi_1+\cdots+\xi_n}{p^{-n}} \to \xi(D)$.

4. Convergence of Series of Independent Random Variables

Theorem 4.1. *Let $\{\xi_n(\omega)\}$ be a sequence of independent RV satisfying the condition:*

there exists a sequence $\{a_n\}, a_n \in \mathbb{R}_+$, such that $|m_{\xi_n}(j)|_p \leq a_n^j$ and $a_n \to 0, n \to \infty$.

Then the series $\sum_{n=1}^{\infty} \xi_n$ converges with respect to a distribution.

Proof. Consider sums $S_n(\omega) = \sum_{k=1}^{n} \xi_k(\omega)$ and their characteristic functions $\varphi_{S_n}(t) = \prod_{k=1}^{n} \varphi_{\xi_k}(t)$. Using the estimate of moments, we get that functions $\varphi_{\xi_n}(t)$ are analytic on $\mathcal{U}_\rho, \rho \in \Gamma_{\mathbb{Q}_p}$ beginning with some $n = n_\rho$. Consequently, there exists such ball $\mathcal{U}_R$ that all $\varphi_{\xi_n} \in \mathcal{A}(\mathcal{U}_R, \mathbb{Q}_p)$. Thus, $\varphi_{\xi_n}, \varphi_{S_n}$ are continued to the functions of the class $\mathcal{A}(\mathcal{W}_R, \mathbb{C}_p)$. It suffices to show $\alpha_{nm} = |||\varphi_{S_m} - \varphi_{S_k}|||_R \to 0$.

Using the proof of Theorem 3.1, we get that $\varphi_{\xi_n}(z) = 1 + A_n(z)$, where $|||A|||_R \to 0$. Thus: $\forall \epsilon > 0 \exists N_\epsilon : |||A|||_R < \epsilon$. In particular, we have that the sequence $\{\varphi_{S_n}\}$ is bounded. Indeed, set $c = 1 + \max_{1\leq n \leq N_1} |||A_n|||_R$. Then:

$$|||\varphi_{S_m}|||_R \leq$$

$$\max(1, \max_{1\leq n\leq m} |||A_n|||_R, \max_{1\leq i\neq j\leq m} |||A_i|||_R \, |||A_j|||_R, \ldots,$$

$$\max_{1\leq i_1\neq\cdots\neq i_k\leq m} |||A_{i_1}|||_R \cdots |||A_{i_k}|||_R, \ldots).$$

Each product is estimated by c^{N_1}.

Further, we have

$$\alpha_{mk} \leq |||\varphi_{S_k}|||_R \, ||| \prod_{n=k+1}^{m} \varphi_{\xi_k} - 1|||_R$$

for $m \geq k$. To finish the proof we choose $k, m \geq N_{\epsilon/c^{N_1}}$.

Example 4.1. Let $\{\xi_n\}$ be independent discrete RV, $\xi_n = u_{n1}, ..., u_{nN_n} \in \mathbb{Q}_p$ with probabilities $\mathbf{P}_{n1}, ..., \mathbf{P}_{nN_n} \in \mathbb{Q}_p$, $\sum_{j=1}^{N_n} \mathbf{P}_{nj} = 1$. If

$$T_{nj} = \max_{1\leq k \leq N_n} |u_{nk}^j \mathbf{P}_{nk}|_p \leq A_n^j,$$

where $a_n \to 0$, then the series of these RV convergence with respect to a distribution.

For example, let $\eta_n = A, B \in \mathbb{Q}_p$ with probabilities $a, b \in \mathbb{Q}_p$, $a+b = 1$. Then the series $\sum_{n=0}^{\infty} \eta_n$ converges in $\mathbb{Q}_p$. The series $\sum_{n=1}^{\infty} n!\eta_n$ converges in $\mathbb{Q}_p$ for all p.

Let $\xi_n = p^{2n}, p^{3n}$ with probabilities $a_n = 1/p^n, b_n = 1 - a_n$. Then $T_{nj} \leq (p^{-2n+1})^j$.

Let $\xi_n = 1, 0$ with probabilities $a_n + 1/p^n, b_n = 1 - a_n$. Then the series $\sum_{n=1}^{\infty} \xi_n$ diverges. Indeed, if $S_n \to S(D)$, then, in particular, $\mathbf{M}S_n \to \mathbf{M}S$. Hence $\mathbf{M}\xi = \sum_{n=1}^{\infty} p^{-n}$. But this series diverges in $\mathbb{Q}_p$.

For the example with RV η_n we get $\mathbf{M} \sum_{n=0}^{\infty} p^n \eta_n = (Aa+Bb)/(1-p)$. In the same way let $\eta_n = 0, 1$ with probabilities $1/2$. Then the series $\sum_{n=1}^{\infty} n^2(n+1)!\eta_n$ converges with respect to a distribution and $\mathbf{M}\eta = 1$. The series $\eta = \sum_{n=1}^{\infty} n^5(n+1)!\eta_n$ also converges and $\mathbf{M}\eta = 13$.

Theorem 4.2. *Let $\{\xi_n\}$ be a sequence of independent RV, $\xi_n \sim N(a_n, b_n)$, where $a_n, b_n \in \mathbb{Q}_p$. Then the series $\xi = \sum_{n=1}^{\infty} \xi_n$ converges with respect to a distribution if and only if the series $a = \sum_{n=1}^{\infty} a_n$ and $b = \sum_{n=1}^{\infty} b_n$ converge. The sum is the Gaussian RV $\xi \sim N(a, b)$.*

Proof. 1. Let series $\sum_{n=1}^{\infty} a_n$ and $\sum_{n=1}^{\infty} b_n$ converge. Introduce RV $\eta_n = \xi_n - a_n$. It suffices to show that the series $\sum_{n=1}^{\infty} \eta_n$ converges. We use the estimate $|m_{\eta_n}(j)|_p \leq |b_n|_p^j$. But $|b_n|_p \to 0$. Thus, the series converges. To show that ξ is a Gaussian RV, it suffices to compute its characteristic function.

2. Now let the series $\xi = \sum_{n=1}^{\infty} \xi_n$ converges. Then $S_n = \sum_{k=1}^{n} \to \xi(D)$ and $\mathbf{M}S_n = \sum_{k=1}^{n} a_k \to \mathbf{M}\xi$. Further, we see that the condition $S_n \to \xi(D)$ implies that $\mathbf{M}S_n^2 \to \mathbf{M}\xi^2$.

5. The p-adic White Noise. An Analogue of Hida's Calculus

The Gaussian distributions on infinite-dimensional spaces play a very significant part in the contemporary probability theory and mathematical physics. The white noise distribution [34] occupies a place of its own in the Gaussian distributions. On one hand, this distribution has a very simple structure (a unit covariant operator in a Hilbert space) and on the other hand the white noise distribution is encountered in many physical problems [35], [36] in the Feynman integration theory inclusive [36]. The foundations for the theory of the non-Archimedean white noise were laid in the report [60].

There is no problem to introduce a p-adic valued white noise distribution inside of the previous axiomatic as an infinite product of Gaussian distributions on $\mathbb{Q}_p$ of the class $N(0,1)$. But it would be much more interesting to introduce the white noise on the basis of L_2-theory. It would be similar to the real case. And we should realize without any problem a p-adic analogue of Hida's theory of white noise functionals in the L_2-representation. Now we begin to construct a p-adic white noise distribution with the aid of the standard scheme [34]. As first step, we define a p-adic white noise distribution on the basis of our theory of Gaussian distribution on infinite-dimensional space using the Laplace-transform definition.

1. Square integrable white noise functionals. Let $K = \mathbb{Q}_p$. We introduce spaces of p-adic sequences

$$\mathcal{S}_k = \{f = (f_n) \in \mathbb{Q}_p^\infty : \lim_{n\to\infty} |f_n|_p\, p^{kn} = 0\},$$

$$||f||_k = \max_n |f_n|_p\, p^{kn}, \qquad k = 0, \pm 1, \ldots;$$

$$\mathcal{S}_\infty = \lim_{k\to\infty} \operatorname{proj} \mathcal{S}_k = \{f \in \mathbb{Q}_p^\infty : \lim_{n\to\infty} |f_n|_p\, p^{kn} = 0,\ k = 1, 2, \ldots\};$$

$$\mathcal{S}_{-\infty} = \lim_{k\to-\infty} \operatorname{ind} \mathcal{S}_k = \{f \in \mathbb{Q}_p^\infty : \text{there exists } k = -1, -2, \ldots,$$

$$\text{such that } \lim_{n\to\infty} |f_n|_p\, p^{kn} = 0\}.$$

The space $\mathcal{S}_\infty$ is the non-Archimedean Fréchet space and $\mathcal{S}_{-\infty}$ is a complete non-Archimedean locally convex space. The equality $\mathcal{S}'_\infty = \mathcal{S}_{-\infty}$ holds true. Just as in Chapter 3, for the spaces $\mathcal{A}(\mathbb{Q}_p^n)$, $\mathcal{A}'(\mathbb{Q}_p^n)$ we introduce functional spaces $\mathcal{A}_0(\mathcal{S}_\infty)$ and $\mathcal{A}(\mathcal{S}_{-\infty})$ and the spaces of distributions $\mathcal{A}'_0(\mathcal{S}_\infty)$ and $\mathcal{A}'(\mathcal{S}_{-\infty})$ and define the Laplace transformation for the distributions $\mu \in \mathcal{A}'_0(\mathcal{S}_\infty)$. The Laplace transformation $L: \mathcal{A}'_0(\mathcal{S}_\infty) \to \mathcal{A}(\mathcal{S}_\infty)$ is an isomorphism, $\mathcal{S}_\infty \subset \mathcal{H} \subset \mathcal{S}_{-\infty}$, $\mathcal{H} \equiv \mathcal{H}_{(1)}$.

Definition 5.1. *The p-adic white noise is a distribution $W \in \mathcal{A}'(\mathcal{S}_{-\infty})$ with the Laplace transformation*

$$L'(W)(f) = \exp\left\{\frac{1}{2}(f,f)\right\}, \qquad (f,f) \equiv (f,f)_\lambda, \qquad \lambda = (1).$$

We consider a system of Hermitian polynomials associated with the distribution of the p-adic white noise:

$$h_n(t) = (-1)^n e^{t^2/2}\frac{d^n}{dt^n}e^{-t^2/2}, \qquad h_\alpha(f) = h_{\alpha_1}(f_1)\cdots h_{\alpha_n}(f_n)\cdots,$$

$$|\alpha| = \sum_{j=1}^{\infty}\alpha_j < \infty.$$

On the space $\mathcal{A}(\mathcal{S}_{-\infty})$ we consider an inner product

$$(F,G) = \int F(f)\bar{G}(f)W(df).$$

The Hermitian polynomials are orthogonal with respect to the inner product and $|h_\alpha|^2 = \alpha!$. Just as in Ch. 2, we cannot normalize these polynomials. The Hermitian polynomials form a basis in Fréchet space $\mathcal{A}(\mathcal{S}_{-\infty}):\ F(f) = \sum_\alpha \tilde{F}_\alpha h_\alpha(f)$. On this space we introduce a non-Archimedean continuous norm $||F|| = \max_\alpha |\tilde{F}_\alpha|_p\sqrt{|\alpha!|_p}$. The completion of the space $\mathcal{A}(\mathcal{S}_{-\infty})$ according to this norm (cf. Ch. 3) is known as the space of functionals of white noise, which are square summable, and is denoted by $L_2(\mathcal{S}_{-\infty}, W)$. This is a non-Archimedean Hilbert space of the type $\lambda = (\alpha!)$.

We set $\Omega = \mathcal{S}_{-\infty}$, $\mathbf{P} = W$. But instead of an algebra of analytic functions, we consider the space $L_2(\mathcal{S}_{-\infty}, W)$ as the space of RV.

The L_2-RV are said to be square summable *functionals of white noise* as Hida's theory of white noise functionals. Since we cannot obviously define any other RV in the framework of the suggested approach, we shall use the term "random variables".

As usual, we introduce the mathematical expectation and the moments of a random variable.

If the linear functional $\mathbf{P}_\xi : f \to \mathbf{M}f(\xi)$ is defined and continuous on the Fréchet space $\mathcal{A}(\mathbb{Q}_p)$, then it is called the distribution of RV ξ. In this case we have

$$\mathbf{M}f(\xi) = \int_{\mathbb{Q}_p} f(x)\mathbf{P}_\xi(\mathrm{d}x).$$

Thus, the situation is different from the axiomatic of generalized probability spaces.

Theorem 5.1. *RV ξ has a distribution if and only if ξ has moments of all orders whose growth is not stronger than*

$$|\mathbf{M}\xi^n|_p \le Cr^n, \qquad C = C_\xi, \qquad r = r_\xi.$$

To prove this theorem, it is sufficient to note that

$$\mathbf{M}f(\xi) = \sum_{n=0}^{\infty} \frac{f^n(0)}{n!}\mathbf{M}\xi^n$$

and use the definition of the topology in $\mathcal{A}(\mathbb{Q}_p)$.

A random vector is any vector $\xi = (\xi_1, \ldots, \xi_n)$ composed of RV. The distribution of random vector is defined if and only if the growth of the moments is not stronger than algebraic.

RV are independent if their joint distribution has the form $\mathbf{P}_\xi = \mathbf{P}_{\xi_1} \times \ldots \times \mathbf{P}_{\xi_n}$. Thus, in the formalism of white noise the concept of independence has sense not for all RV. There exist such RV for which we cannot discuss a problem of dependence or independence. Let ξ be RV having a distribution $\mathbf{P}_\xi$. Then, in a certain neighborhood of zero, the characteristic function of RV is defined in the same way as in section 1. Theorems 1.1–1.4 and proposition 1.4 are also valid in this case.

L_2-convergence of Gaussian series and white noise functionals. The mappings $\bar{\xi}_n : \Omega \to \mathbb{Q}_p$,

$$\bar{\xi}_n(\omega) = (e_n, \omega), \qquad n = 1, 2, \dots,$$

where $\{e_n\}$ is a canonical basis in $\mathcal{H}$, can serve as examples of independent Gaussian variables which have distribution $N(0, 1)$.

Remark 5.1. *We can use the Gaussian variable $\xi \sim N(0,1)$ to obtain a Gaussian variable $\eta \sim N(m, \sigma^2)$ for any number $\sigma \in \mathbb{Q}_p$. However, generally speaking, there we cannot obtain the Gaussian RV $\eta \sim N(m, b)$ even in the case when b is a positive rational number.*

Theorem 5.2. *L_2-limit of the sequence $\{\eta_n(\omega)\}$ of Gaussian RV is also a Gaussian RV.*

To prove this theorem, we must consider an explicit representation of RV as functionals of white noise.

Probably it would be more useful to investigate the questions proposed in section 1 inside of L_2-formalism.

Let us consider a sequence of independent similarly distributed Gaussian RV $\xi_n(\omega) \sim N(0,1)$. We have an analogue of Theorem 4.2 for the L_2-convergence.

Theorem 5.3. *Let the sequence of p-adic numbers $a_n \to 0, n \to \infty$. Then the series $\sum_{n=1}^{\infty} a_n \xi_n(\omega)$ converges to the Gaussian random variable*

$$\eta(\omega) \sim N(0, \sum_{n=1}^{\infty} a_n^2).$$

For instance, the following series of independent standard Gaussian RV converges over the field of p-adic numbers:

$$\eta(\omega) = \sum_{n=1}^{\infty} p^{n-1} \xi_n(\omega) \sim N(0, 1/(1-p^2));$$

$$\eta(\omega) = \sum_{n=1}^{\infty} n! \xi_n(\omega) \sim N(0, \sum_{n=1}^{\infty} (n!)^2).$$

Theorem 5.4. *Let the sequence of p-adic numbers $a_n \to 0$, $n \to \infty$. Then the series $\eta(\omega) = \sum_{n=1}^{\infty} a_n (\xi_n^2(\omega) - 1)$ converges.*

For instance, the series

$$\eta(\omega) = \sum_{n=1}^{\infty} n!(\xi_n^2(\omega) - 1)$$

converges.

In the ordinary probability theory the series considered above can be realized on the mathematical level of rigorous in the framework of Hida's calculus as generalized Brownian functionals. And here they are ordinary random variables.

3. Generalized white noise functionals. A theory of p-adic valued generalized white noise functionals was developed in [60]. There is a possibility to realise as p-adic valued generalised white noise functionals such formal series of Hermitian polynomials with rational coefficients, which are not well defined just in Hida's distributions sense.

Let us consider the nested Hilbert space

$$\mathcal{A}(\mathcal{S}_{-\infty}) \subset L_2(\mathcal{S}_{-\infty}, W) \subset L_2'(\mathcal{S}_{-\infty}, W) \subset \mathcal{A}(\mathcal{S}_{-\infty})'.$$

Functions $G(\omega) \in L_2'(\mathcal{S}_{-\infty}, W)$ are said to be L_2-generalised white noise functionals and functions $G(\omega) \in \mathcal{A}(\mathcal{S}_{-\infty})$ are said to be generalized white noise functionals.

We can construct a great number of examples of p-adic valued generalised white noise functionals, which cannot be realized as Hida's functionals (all coefficients are rational number and formal expressions are invariant with respect to a number field).

For example, let $p = 2$ and

$$G(\omega) = \sum_{n=0}^{\infty} 3^{n!}/n! h_n(\omega_n).$$

This series does not converge in $L_2(\mathcal{S}_{-\infty}, W)$ and already in $L_2'(\mathcal{S}_{-\infty}, W)$. But it converges in $\mathcal{A}(\mathcal{S}_{-\infty})'$ and $G(\omega)$ is a well defined p-adic valued generalized white noise functional. If we realize this series as a series of standard real white noise functionals, then this series diverges in the sense of Hida's distributions.

Thus, we could also consider the p-adic valued white noise formalism as a possibility to regulate divergent stochastic expressions (with rational coefficients) of standard real theory of probability, based on the Kolmogorov axiomatic. The previous example was created to illustrate the situation on the basis of "pure" generalised white noise functional. But we can present examples of L_2-functional, which have divergent real realization (already in the sense of Hida's distribution's). For instance,

$$G(\omega) = \sum_{n=0}^{\infty} n!^{n!} h_n(\omega_n).$$

This function is the L_2 white noise functional for all p.

Of course, our regularization can be applied to only very special (but sufficiently extended) class of Hermitian coefficients.

4. Stochastic processes on the basis of white noise. Let T be a subset of the field $\mathbb{Q}_p$. The collection of RV $X = \{\xi_t\}_{t\in T}$ is called a stochastic process with a time p-adic interval T. Note that in our formalism

$$X : T \to L_2(\mathcal{S}_{-\infty}, W(df)), \tag{5.1}$$

i.e., for any $t \in T$ the random variable $\xi_t(\omega)$ is a square summable functional of the p-adic white noise.

The stochastic process X is said to be a process with continuous (in the quadratic mean) paths if mapping (5.1) is continuous.

Theorem 5.5. *Let $T = Z_p$. Then every stochastic process $X = \{\xi_t\}_{t\in T}$ with continuous paths is uniquely defined by its values at the time moments $t = 0, 1, 2, \ldots$, i.e., the random sequence $\{\xi_n\}_{n=0}^{\infty}$ uniquely defines the stochastic process.*

To prove this theorem, it is sufficient to note that the set of natural numbers is everywhere dense in the ring of p-adic integers.

Here is a simple example of a process with paths continuous in the quadratic mean, $T = Z_p$.

Suppose that the sequence $\{\xi_n(\omega)\}$ consists of independent Gaussian RV with the distribution $N(0, 1)$. Note that every $t \in Z_p$ can be expanded in the series

$$t = t_0 + t_1 p + \ldots + t_j p^j + \ldots,$$

where $t_j = 0, 1, \dots, p-1$. We set

$$\eta_t(\omega) = t_0\xi_0(\omega) + t_1 p\xi_1(\omega) + t_2 p^2 \xi_2(\omega) + \dots$$

By virtue of Theorem 5.3, this series converges in the quadratic mean for any $t \in Z_p$ to the Gaussian random variable

$$\eta_t(\omega) \sim N(0, \sum_{j=0}^{\infty} t_j^2 p^{2j}).$$

We shall show that the stochastic process $\eta_t(\omega)$ is continuous in the quadratic mean. Note that

$$||\eta_t - \eta_s||_2 = \max_{0 \le j < \infty} |t_j - s_j|_p p^{-j}.$$

Let $|t-s|_p < p^{-k}$. Then $t_j = s_j$, $j = 0, 1, \dots, k-1$, i.e., $||\eta_t - \eta_s||_2 \le p^{-k} \to 0$, $k \to \infty$. Thus $X = \{\eta_t(\omega)\}$ is a Gaussian stochastic process with paths continuous in the quadratic mean.

Let $w(t, \omega)$ be an ordinary Wiener process over the field of real numbers. Then $w(t, \omega) \in L_2(\Omega, \mathbf{P}(d\omega))$ for any t. This is not so in the p-adic case. We can prove this in many ways. We shall use the representations of Paley–Wiener (see, for instance, [34, p. 79]) for the Wiener process $w(t, \omega)$ and take, in the p-adic case, this representation as one of the realizations of the Wiener process $w(t, \omega)$.

Let $\{\xi_k(\omega)\}$ and $\{\eta_k(\omega)\}$ be two independent sequences of independent Gaussian random variables with a normal distribution $N(0, 1)$, chosen from the standard sequence $\{\bar{\xi}_n(\omega)\}$. Let us consider a series of Gaussian random variables

$$w(t, \omega) = \frac{1}{\sqrt{2\pi}} \Big\{ t\xi_0(\omega) + \sum_{n \neq 0} (\xi_n(\omega) \sin nt + \eta_n(\omega)(\cos nt - 1))/n \Big\}, \tag{5.2}$$

where, in the p-adic case ($p \neq 2$), the number π is defined as $\Gamma_p^2(1/2)$, where $\Gamma_p(x)$ is a p-adic Γ-function of Morita (see [106, p. 108]).

In a real case, series (5.2) converges in the quadratic mean and in the p-adic case this series diverges in the quadratic mean for arbitrarily small moments t. Thus, we have an important open problem to build p-adic valued Wiener process.

Chapter IX

p-Adic Superanalysis

The next natural step in the development of the p-adic physics is the consideration of p-adic fermions and supersymmetric theories. Some results in this direction were proposed by V. Vladimirov and I. Volovich in [125]. They considered the superspace over an arbitrary locally compact ultrametric number field and proposed a number of results about analysis on such superspace. The main part of this paper was devoted to real superanalysis and the non-Archimedean part was a brief remark about the possibility to generalize the real superanalysis to the non-Archimedean case.

Investigations in this subject were continued in the series of papers of R. Cianci and A. Khrennikov [18]–[20]. At first, we have seen that some of propositions in the paper of Vladimirov and Volovich are not true (see [66], [19]). For example, we have constructed [19] an example of a function $f(\theta)$, where θ is an odd variable, which is infinitely superdifferentiable but it isn't a polynomial. The next problem in p-adic case is the inverse function theorem. This theorem isn't true for superdifferential functions and we must introduce a notion of strictly superdifferential function to resolve this problem (there are very useful old results of N. Bourbaki, see applications of Bourbaki's theory in [19]).

In this Chapter we begin to construct a mathematically p-adic superanalysis. We shall use the functional differential geometry approach to superanalysis proposed by B. De Witt, A. Rogers, A. Jadczyk and K. Pilch, V. Vladimirov and I. Volovich (see references in [18]–[19], [66], [56]). The functional superanalysis is also called a differential geometry approach to superanalysis in some papers (for instance, in papers of A. Rogers). According to this approach we consider, first of all, a set of superpoints (superspace) and then we generate a suitable sheaf of functions on this set of points. It could be compared with an algebraic approach of J. Martin, F. Berezin, B. Kostant D. Leites (see references and discussion in [66]), where "functions" of anticommuting "variables" are defined as elements of Grassmann algebras.

As the first step we construct a series of infinitely generated p-adic Banach superalgebras $B_{p,k}$, $k \in \mathbf{Z}$, and study the spectrum properties of these superalgebras. The main result is the zero spectrum of every soul element of $B_{p,k}$ (these algebras play in p-adic superanalysis the role of A. Roger's Banach superalgebra $A = l_1(\wedge^\infty \mathbb{R})$). Then we consider the projective limit $B_{p,\infty} = \lim_{k\to\infty} \operatorname{proj} B_{p,k}$ and inductive limit $B_{p,-\infty} = \lim_{k\to-\infty} \operatorname{ind} B_{p,k}$ of these algebras and study topological and spectral properties of p-adic locally convex superalgebras $B_{p,\infty}$ and $B_{p,-\infty}$. This rich theory of infinitely generated Banach and locally convex superalgebras is specific of the p-adic case. The real analogues of these algebras will be only Banach spaces but not Banach algebras.

We also study matrix algebras over $B_{p,k}$ and $B_{p,\infty}$, $B_{p,-\infty}$ and, by using the spectral properties of soul matrices, we define an exponent function on some additive subgroup of the matrix algebra (it is impossible to define $\exp\{U\}$ for an arbitrary supermatrix in the p-adic case).

Then we shall study the properties of superdifferential, strictly superdifferential and superanalytic functions and, in particular, the connection between this functional classes.

On the base of this superdifferential calculus we shall develop the theory of p-adic supermanifolds over infinitely generated p-adic Banach superalgebras. We'll pay our main attention to supersmooth manifolds which are nonsuperanalytic in pure anticommuting sector. The next interesting moment is the infinite dimension of these supermanifolds.

It is well known that every finite dimensional paracompact ultrametric manifold can be represented as the union of the disjoint balls but in our case we have more interesting geometrical objects. We hope that the theory of p-adic supermanifolds will be a good base for the p-adic supergravity and superstring theory.

The theory proposed is the theory of p-adic valued fermions (p-adic valued p-adic physics).

1. Infinitely Generated p-adic Banach Superalgebras

We recall here the following definition

Definition 1.1 *A supercommutative non-Archimedean Banach superalgebra over* $\mathbb{Q}_p$ *is a* $\mathbf{Z}_2$ *graded commutative Banach algebra over* $\mathbb{Q}_p : \Lambda = \Lambda_0 \bigoplus \Lambda_1$ *such that the norm is a non-Archimedean:* $\|x+y\| \leq \max \|x\|, \|y\|$. *Moreover, we require that*

$$\|x+\theta\| = max(\|x\|, \|\theta\|)$$

for $x \in \Lambda_0, \theta \in \Lambda_1$.

A simple example of supercommutative non-Archimedean Banach superalgebras is the exterior algebras $\wedge^n \mathbb{Q}_p$ endowed with max-norm.

Now we construct the example of an infinite-dimensional non-Archimedean superalgabra which will be very useful in applications.

Consider an infinite number of anticommuting elements (*the generators*) $\{q_j\}_{j=1}^{\infty} : q_i q_j = -q_j q_i$.

By means of this set, we construct the exterior p-algebra $\wedge^\infty \mathbb{Q}_p$ of formal series.

Let us denote with M_∞ the set of sequences of positive integer: $\mu = (j_1, ..., j_k)$ with $1 \leq j_1 < j_2 ... < j_k$ ($\emptyset$ denotes the empty sequence) and define, for each $\mu \in M_L$,

$$\beta_\mu = q_{j_1} \dots q_{j_k} \text{ and } \beta_\emptyset = 1 \tag{1.1}$$

The products $\beta_\mu \beta_\nu$ allows us to define the terms $f^\rho_{\mu\nu}$ by means of equation $\beta_\mu \beta_\nu = f^\rho_{\mu\nu} \beta_\rho$. Now,

$$f^\rho_{\mu\nu} = \text{degr}\,(\Pi) \text{ if } \ \mu \cup \nu = \rho \text{ and } 0 \text{ otherwise.}$$

where Π denotes the permutation which sends $\mu \cup \nu \mapsto \rho$.

We define

$$f^\rho_{\mu\nu} = \delta^\rho_{\mu\cup\nu} = \sigma(\mu - \nu, \nu)$$

where $\mu - \nu$ is the sequence obtained by μ by deleting the terms of the sequence ν and $\text{degr}(\mu) \equiv |\mu| = 0$ if k is even and 1 if k is odd; we define also $\|\mu\| = j_1 + j_2 + ...$ which satisfies $\|\rho\| = \|\mu\| + \|\nu\|$.

The following equalities can be useful to the reader which wishes to perform our calculations:

$$f^\rho_{\mu\nu} = f^\rho_{\nu\mu}(-1)^{|\mu\nu|}$$

$$f^\rho_{\nu\mu} f^\gamma_{\alpha\rho} = f^\rho_{\alpha\nu} f^\gamma_{\rho\mu} \sigma(\mu, \nu)\sigma(\mu + \lambda, \nu) =$$

$$\sigma(\mu, \nu + \lambda)\sigma(\nu, \mu)\beta_\mu = \sigma(\mu, \nu)\beta_{\mu-\nu} \wedge \beta_\nu$$

Now we construct an infinite series of p-adic Banach superalgebras.

More precisely, for every integer k we define the supercommutative Banach algebras

$$B_{p,k} \equiv c_k\left(\wedge^\infty \mathbb{Q}_p\right) = \left\{ q = \sum_{\mu \in M_\infty} b_\mu \beta_\mu : b_\mu \in \mathbb{Q}_p, \lim_{\|\mu\| \to \infty} p^{k\|\mu\|} |b_\mu|_p = 0 \right\} \tag{1.2}$$

with the natural $\mathbf{Z}_2$-graded structure $B^0_{p,k} \oplus B^1_{p,k}$.

The non-Archimedean norm in $B_{p,k}$ is defined as

$$\|q\|_{p,k} = \max_\mu \left\{ p^{k\|\mu\|} |b_\mu|_p \right\}.$$

This norm is a $\mathbf{Z}_2$-graded one.

Proposition 1.1 *The algebras $B_{p,k}$ are non-Archimedean Banach superalgebras with basis $\{\beta_\mu\}$.*

Proof. It is evident that the norm $\|\cdot\|_{p,k}$ is a non-Archimedean one:

$$\|q + q'\|_{p,k} \leq \max(\|q\|_{p,k}, \|q'\|_{p,k}).$$

Let us prove that for every $q, q' \in B_{p,k}$ the element $qq' \in B_{p,k}$. By setting $q = \sum_\nu c^\nu \beta_\nu$, $q' = \sum_\mu c'^\mu \beta_\mu$, we get:

$$a = qq' = \sum_\mu \sum_\nu \sum_\rho c^\nu c'^\mu f^\rho_{\nu\mu} \beta_\rho$$

If $q, q' \in B_{p,k}$, for every $\varepsilon > 0$ there exists a N such that, if $\|\mu\| > N$ then $|c_\mu|_p < p^{-k\|\mu\|}\varepsilon$ and $|c'_\nu|_p < p^{-k\|\nu\|}\varepsilon$. Now, by using the strong triangular inequality, we get

$$|a_\rho|_p \leq \max_{|\mu||\nu|,\rho=\mu\cup\nu,\mu\cap\nu=\emptyset} |c_\mu|_p |c'_\nu|_p \tag{1.3}$$

Let $\|\rho\| > 2N$; then $\|\mu\| > N$ or $\|\nu\| > N$ and $|c_\mu|_p < p^{-k\|\mu\|}\varepsilon$ or $|c'_\nu|_p < p^{-k\|\nu\|}\varepsilon$. Supposing that $\|\mu\| > N$, We get

$$|a^\rho|_p p^{-k\|\rho\|} < p^{-k\|\rho\|}\varepsilon \max_{\mu\nu}(p^{-k\|\mu\|}|c'_\nu|_p)$$

$$< \varepsilon \max_\nu(p^{k\|\nu\|}|c'_\nu|_p) < \varepsilon \|q'\|_{p,k}$$

so $\lim_{\|\mu\|\to\infty} |a_\rho|_p p^{-k\|\rho\|} = 0$. We get also the Banach algebra inequality:

$$\|qq'\|_{p,k} \leq \|q\|_{p,k}\|q'\|_{p,k}.$$

The system of elements $\{\beta_\mu\}$ is a Banach space basis because

$$\|q - \sum_{\nu>N} c_\nu \beta_\nu\|_{p,k} \to 0, \text{whenever } , N \to \infty.$$

This set is an orthogonal basis in the sense of non-Archimedean Banach spaces .

Now we observe that, for all k, $B_{p,k} \subset B_{p,k-1}$.

For this reason we consider the chain of algebras

$$....B_{p,k} \subset B_{p,k-1} \subset .... \subset B_{p,0} \subset B_{p,-1} \subset ....$$

Now we define the linear space

$$B_{p,\infty} \equiv \bigcap_{k=-\infty}^{+\infty} B_{p,k}, \qquad B_{p,-\infty} \equiv \bigcup_{k=-\infty}^{+\infty} B_{p,k}$$

The first space obtained with a projective limit is endowed with the projective topology; the second one obtained with a inductive limit is endowed with the inductive topology.

Proposition 1.2 *$B_{p,\infty}$ and $B_{p,-\infty}$ are locally convex topological non-Archimedean superalgebras with basis $\{\beta_m u\}$; moreover $B_{p,\infty}$ is a Frechèt superalgebras (complete ultrametric algebra) while $B_{p,-\infty}$ is dual Frechèt superalgebra.*

We observe that equation $B_{p,-\infty} = B'_{p,\infty}$ (the dual) was used to show the last fact.

2. Spectral Properties

In this section we study spectral properties of the algebras constructed in the previous sections.

To this end we define the *body* map $\varepsilon: B_{p,k} \to \mathbb{Q}_p$; if we set $x = \sum_\mu x^\mu \beta_\mu$, then $\varepsilon(x) = x^\phi$. In the same way we define also the *soul* map $s: B_{p,k} \to B_{p,k} - \mathbb{Q}_p$ as: $s(x) = x - \varepsilon(x)$. The same definition is applied to algebras $B_{p,\infty}$ and$B_{p,-\infty}$.

Theorem 2.1. *For each $c \in B_{p,k}$, $\varepsilon(c) = 0$ and for every real θ : $0 < \theta < 1$ there exists a real $\lambda : 0 < \lambda < 1$ such that $\|c^n\|_{p,k} \leq \lambda\theta^n$ for sufficiently large n.*

Proof. Suppose that c is even. By using the expansion of c on the basis $\{e_\mu\}$, for every real $\varepsilon > 0$, we can write $c = a + b$, where a is nilpotent (e.g. there is a number $N(\varepsilon)$ for which $a^N = 0$) and $\|b\|_{p,k} < \varepsilon$.

By using this equation, and for every $n > N$ we get

$$\|c^n\|_{p,k} = \|\sum_{k=0}^{N} \binom{n}{k} a^k b^{n-k}\|_{p,k} \leq \max_{0 \leq k \leq N} |\binom{n}{k}|_p \varepsilon^{n-k}[\|a\|_{p,k}]^k$$

$$\leq \max_{0 \leq k \leq N} \varepsilon^{n-k}[\|a\|_{p,k}]^k$$

Since $\binom{n}{k}$ are natural numbers $|\binom{n}{k}|_p \leq 1$ for every n, k.

If we denote with $K = \max_{0 \leq k \leq N}[\|a\|_{p,k}]^k$, we have $\|c^n\|_{p,k} \leq k/\varepsilon^N \varepsilon^n$.

Take $\varepsilon = \gamma A$ where $0 < \gamma < 1$. Then $\|c^n\|_{p,k} \leq k\gamma^n/\gamma^N \theta^n$ where $\lambda_n = k\gamma^n/\gamma^N \to 0$ when $N \to 0$. The proof of this theorem when c has no definite parity is similar.

As a consequence of this theorem, every $c \in s(B_{p,k})$ has zero spectrum.

Proposition 2.1 *Let $c \in B_{p,k}$; the element c^{-1} exists if and only if $c_0 = \varepsilon(c) \neq 0$. In this case we have:*

$$c^{-1} = \sum_{n=0}^{\infty}(-1)^n \left(\frac{c - c_0}{c_0}\right)^n$$

The expansion gives formally the correct value of c^{-1}; on the other hand, the convergence is ensured by the previous theorem.

As a consequence of this proposition, every $c \in s(B_{p,k})$ has zero spectrum.

The same result is also true for algebras $B_{p,\infty}$ and $B_{p,-\infty}$ as a consequence of the projective and inductive topologies.

Now consider the algebra $gl_{p,k}(n)$ of $(n \times n)$ matrices with elements in $B_{p,k}$ and $GL_{p,k}(n)$ the group (at this stage it is only an algebraic group) of non-singular $gl_{p,k}(n)$ matrices.

Theorem 2.2. *A matrix $A \in gl_{p,k}(n)$ is singular if and only if* $\det \varepsilon(A) = 0$.

We consider, first of all the case in which the elements of A are even.

Proposition 2.3. *Let A even and $\varepsilon(A) = 0$. For every real θ : $0 < \theta < 1$, there is a real $\lambda : 0 < \lambda < 1$ such that $\|A^n\|_{p,k} \leq \lambda\theta^n$ for sufficiently large n.*

Proof. Since every element $x \in B_{p,k}$ can be written in the form $x = y + z$ where y is sum of strictly nilpotent terms and $|z| \leq \varepsilon$ we apply this result to the elements of $A = Y + Z \in gl_{p,k}(n)$. This means that for each i, j there exists a positive number $N(i,j)$ so that $(Y_i^j)^{N(i,j)} = 0$ Taking $N = n^2 \times \mathrm{Max}_{i,j}\,\{N(i,j)\}$ for $i, j = 1, ..., n$, we deduce that, for each integer $s \geq 0$, all the terms containing at least N Y's in the expansion $A^{N+s} = (Y + Z)^{N+s}$ vanish independently of the order in which the Y's appear. We have

$$A^{N+s} = Y^{N-1}Z^{s+1} + Y^{N-2}Z^{s+1}Y + \dots \quad \binom{N+s}{N-1} \text{ terms} +$$

$$Y^{N-2}Z^{s+2} + Y^{N-3}Z^{s+1}YZ + \dots \quad \binom{N+s}{N-2} \text{ terms} + \dots$$

and

$$\|A^{N+s}\|_{p,k} \leq \max\left\{\|Z^{N+s}\|_{p,k}, \dots, |\binom{N+s}{N-1}|_p \|Y^{N-1}\|_{p,k}\|Z^{s+1}\|_{p,k}\right\}$$

The use Theorem 2.1 to complete the proof.

Proposition 2.4. *An even matrix A is singular if and only if* $\det \varepsilon(A) = 0$.

Let $\varepsilon(A) = B$ and consider the expression

$$A^{-1} = B^{-1}\sum_{n=0}^{\infty}(-1)^n \left(\frac{A-B}{B}\right)^n$$

This expansion is convergent for the previous proposition. The inverse implication is trivial since $\varepsilon(A)\varepsilon(A^{-1}) = \mathbf{I}$.

To prove Theorem 2.2, it suffices to expand the matrix $A \in gl_{p,k}(n)$ in the form $A = A_0 + A_1$ by splitting its term in the even and odd part respectively.

Using again the properties of projective and inductive topologies we can prove the same result for the locally convex matrix groups $gl_{p,\infty}$ and $gl_{p,-\infty}$ with the elements belong to $B_{p,\infty}$ and $B_{p,-\infty}$ respectively.

Now we wish to define the exponential and the logarithm of a $gl_{p,k}(n)$ matrix.

We define the maps exp and log of p-adic supermatrices $X \in gl_{p,k}(n)$ by means of the expansions:

$$expX = \sum_{n=0}^{\infty}\frac{X^n}{n!}, \; logX = \sum_{n=1}^{\infty}\frac{X^n}{n}(-1)^{n+1}$$

A main difference with respect to standard superanalysis is the non-convergence of these expansion for *every* matrices.

Actually, we have to fix our attention for the additive subgroup

$$W_\omega \equiv \{X \in gl_{p,k}(n) : \|\varepsilon\,(X)\|_{p,k} < \omega\}$$

where $\omega = p^{1/(1-p)}$.

This subgroup is closed and in the same time open in $gl_{p,k}(n)$. Specialists of non-Archimedean analysis call such objects *clopen* subgroups.

The set $V_\omega \equiv \{\mathbf{1} + W_\omega\}$ is a clopen multiplicative subgroup of $gl_{p,k}(n)$ too.

Theorem 2.3. *The map* exp: $W_\omega \to V_\omega$ *is an isomorphism of p-adic Banach groups; the logarithm is its inverse map.*

Proof. At first, let us prove that the expansion for these functions are convergent on their domains respectively. Set $\varepsilon\,(X) = Y$ and $s(X) = Z$. Then

$$\|X^n/n!\|_{p,k} \leq |n!|_p^{-1} \max_{0\leq k\leq n} |\binom{n}{k}|_p \|Y\|_{p,k}{}^k \|Z^{n-k}\|_{p,k}$$

Let $\|y\|_{p,k} = \gamma\omega$, where $0 \leq \gamma < 1$, and let us take $\theta = \gamma\omega$ for the standard estimate of $\|Z^m\|_{p,k}$.

Now we get:

$$\|X^n/n!\|_{p,k} \leq \lambda\gamma^n(\omega^{-1})^n\omega^n = \lambda\gamma^n \to 0, n \to \infty.$$

Now let us prove that $\exp\,(W_\omega) \subset V_\omega$:

$$\|\varepsilon\,(\exp X) - 1\|_{p,k} = \|\sum_{n=1}^{\infty} \varepsilon\,(X^n)\,/n!\|_{p,k} <$$

$$max_{1\leq n\leq\infty}\left(\omega^n\omega^{(1-n)}\right) = \omega.$$

In the same way we can prove that $\log\,(V_\omega)) \subset W_\omega$.

Let us consider another example of the infinitely generated p-adic Banach superalgebra. It is the algebra of formal expansions

$$l_\infty\left(\wedge^\infty\mathbb{Q}_p\right) = \left\{q = \sum_{\mu\in M_\infty} b_\mu\beta_\mu : b_\mu \in \mathbb{Q}_p, \|q\|_{p,0} = \sup_{\mu\in M_\infty} |b_\mu|_p < \infty\right\},$$

endowed with sup-norm $\|\cdot\|_{p,0}$. It is a non-Archimedean Banach superalgebra and $B_p = c_0\left(\wedge^\infty \mathbb{Q}_p\right)$ is its Banach subsuperalgebra.

This algebra has not such good properties as B_p. For example, Theorem 1.1 is not valid for $l_\infty\left(\wedge^\infty \mathbb{Q}_p\right)$. Now we construct the corresponding example. Let

$$q = q_1q_2 + q_3q_4 + q_5q_6 + \ldots$$

Then $q^n = n! q_1q_2q_3q_4 \ldots q_{2n-1}q_{2n} + \ldots$ and so

$$\|q^n\|_{p,0} \geq |n!|_p$$

In this case we have the following equation for the soul elements whose *quantum* implications are not yet completely clarified:

$$\overline{\lim}_{n\to\infty} \sqrt[n]{\|q^n\|_{p,0}} \geq p^{-1/(p-1)}$$

This equation is not true for Rogers's algebras $l_1(\wedge^\infty(\mathbb{R})$, for Jadczik and Pilch BG algebras and for our B_p algebras. This is the motivation for which, in these case, it was possible to show that an element is not singular if and only if the body is not zero, and, by using a different language, a soul perturbation of classical, real state, does not give origin to any physical quantum effects. In the case of $l_\infty(\wedge^\infty \mathbb{Q}_p)$ algebras this is no longer true; probably one can consider soul perturbations as quantum physical effects.

The condition of the zero limit of the coefficients, as stated in equation (1.2) is the *weaker* condition ensuring the convergence of the expansions there involved. If one try to consider a direct analogous of Rogers algebras $l_1(\wedge^\infty \mathbb{R})$ in p-adic case, so to consider the algebra

$$l_1(\wedge^\infty \mathbb{Q}_p) = \left\{ q = \sum_\mu b_\mu \beta_\mu : b_\mu \in \mathbb{Q}_p, \|q\|_{p,1} = \sum_\mu |b_\mu|_p < \infty \right\}$$

then this algebras is not a non-Archimedean one.

Definition 2.1. *An (n,m)-dimensional p-adic superspace is a $\mathbb{Q}_p$-linear Banach space $\mathbb{Q}_p^{n,m} = \Lambda_0^n \times \Lambda_1^m$ endowed with the norm $\|u\| = \max_{1\leq j\leq n+m} \|u_j\|$.*

$\mathbb{Q}_p^{n,m}$ is also a Λ_0-module: $(au)_j = au_j$ for $a \in \Lambda_0$, $u \in \mathbb{Q}_p^{n,m}$.

We can also define the $(n+m)$-dimensional p-adic superspace as the Banach Λ module $\mathbb{Q}_p^{(n+m)} = \Lambda^{n+m}$ endowed with the usual norm.

Let V and W be p-adic superspaces of dimension $(m+n)$ and $(p+q)$ respectively.

Definition 2.2. *The Banach space of the Λ-left-linear operators from V to W is called $L_p(V, W)$.*

3. Differential Calculus on the p-adic Banach Superspace

Here we study the differential calculus and to the theory of manifolds over infinitely generated p-adic Banach superalgebras.

Great attention will be devoted to the study of the classes of differential maps and *strictly* differential maps; actually, when considering maps over p-adic number field, one gets that some "good" properties like the inverse function theorem and the Schwartz properties for second derivatives can be proved only for *strictly* differential maps.

Our theory of differential maps on p-adic Banach superspace will keep into account a fact which was discussed in Chapter 1. There are maps which have everywhere vanishing derivative and, however, are not constant. This "strange" fact can be related to the non-existence of an order relation in $\mathbb{Q}_p$. An important consequence of this statement is our theorem stating that superfield expansion is no longer true: we shall show an example of odd variables map which is not a polynomial.

Now we study functions on $\mathbb{Q}_p^{n,m}$.

Definition 3.1. *A map $f: \mathbb{Q}_p^{n,m} \to \Lambda$ is called an S-differential map (in the p-adic sense) if it is differentiable in the Frechèt sense and if the Frechèt differential $Df: \mathbb{Q}_p^{n,m} \to \Lambda$ is a multiplicative operator:*

$$Df(h) = \sum_{i=1}^{n+m} h^i (Df)_i$$

for every $h \in \mathbb{Q}_p^{n,m}$.

Any S-differential map f admits the expansion formula

$$f(u+h) = f(u) + \sum_{i=1}^{n+m} h^i (Df)_i + \mathrm{o}(h)$$

Notice that this definition does not imply at all that the definition of the multiplicative operator describing the Frechèt operator is unique. Actually, it is true only if the algebra Λ has a trivial Λ_1-annihilator.

More precisely, if we define the right annihilator $\Lambda_1^{\perp} = \{a \in \Lambda : a\Lambda_1 = 0\}$, we have to require $\Lambda_1^{\perp} = 0$. This is clearly a restriction on the algebras which we are using.

Theorem 3.1. *If Λ is an algebra with nontrivial multiplication operation for odd elements and trivial Λ_1-annihilator then Λ_1 is an infinite-dimensional space.*

In the following of this paper we shall assume that this requirement is always verified; however, many results are true also in the general case.

A concrete example of algebras with trivial Λ_1-annihilator is given by the algebras $B_{p,k}$. The property of trivial Λ_1-annihilator just discussed is true also by considering the tensor product of $B_{p,k}$ with any arbitrary non-Archimedean commutative Banach algebra $E : E \bigotimes B_{p,k}$; the problem of the completion of this space will not be studied here.

In this case the requirement of S-differentiability can be expressed in an explicit way. More precisely, set $x^i \in \mathbb{Q}_p^{n,m} : x^i = \sum_\mu x^{i\mu}\beta_\mu$ and consider the equation

$$f(x^{i\mu} + h^{i\mu}) = f(x^{i\mu}) + \sum_\nu h^{i\nu} \frac{\partial f}{\partial x^{i\nu}} + \ \mathrm{o}\ (h)$$

Now f is S-differentiable if and only if there are functions $D_i f(x)$ such that $\beta_\nu D_i f(x) = \frac{\partial f}{\partial x^{i\nu}}$. In particular, $D_i f(x) = \frac{\partial f}{\partial x^i \emptyset}$.

Definition 3.2. *A map $f\colon U \subset \mathbb{Q}_p^{n,m} \to \Lambda$ is called a superanalytic function if for all $x \in U$ there is an open neighborhood $N \subset U$ of x such that, for all $q \in N$, by setting $r = (q-x)$, we have:*

$$f(q) = \sum_{i_1=0,\ldots,i_{n+m}=0}^{\infty} r^{i_1} \ldots r^{i_{n+m}}\ a_{i_1 \ldots i_{n+m}}$$

for some coefficients $a_{i_1 \ldots i_{n+m}} \in \Lambda$.

We recall that a Frechèt differentiable map $f: E \to F$ (here E, F are Banach spaces) is called strictly derivable in x_0 if,

$$f(y) - f(z) = Df(x_0)(y - z) \bmod o(\|y - z\|)$$

for all $(y, z) \to (x_0, x_0)$ in $E \times E$.

Definition 3.3. *If C^k denote the class of standard k-times continuously Frechèt differentiable maps, denote with $\mathcal{D}^k \subset C^k$ the class of k-times strictly derivable maps.*

Moreover, denote with S^k the class of k-times continuously superdifferential maps and with S^ω the class of S-analytic functions. Finally, denote with $\mathcal{SD}^k = \mathcal{D}^k \cap S^k$ the class of k-times strictly derivable superdifferential maps.

We shall prove now a proposition on classical analysis which we have not found in the literature.

Proposition 3.1 *Let $f \in \mathcal{D}^2$ map from a p-adic Banach space to a further p-adic Banach space; then the second Frechèt differential is symmetric.*

Outline of the proof. Consider the following second order difference

$$\delta_{h,v} = [f(x+h, y+v) - f(x, y+v)] - [f(x+h, y) - f(x, y)]$$

for elements $x, y, x+h, y+v$ in an open neighborhood. By using derivability of f and of its first order differential, one can expand $\delta_{h,v}$ in two different ways by introducing the symbols $D_{xy}f$ and $D_{yx}f$; now, strictly derivability of f and Df, implying that the various rests which are involved are uniform, allows us to complete the proof.

Proposition 3.2 *Let $\mathbb{Q}_p^{n,m} = (B_{p,k}^0)^n \times (B_{p,k}^1)^m$. Let $f: \mathbb{Q}_p^{n,m} \to B_{p,k}$ be S^1 and $\mathcal{D}^k$. Then f is $\mathcal{SD}^k$.*

Proof. We shall prove this point for second derivatives. Since the Frechèt differential is symmetric, for $h, v \in \mathbb{Q}_p^{n,m}$, we have

$$((D^2 f(x), (h, v)) = \sum_{\nu\mu} v^{i_\mu} h^{j_\nu} \frac{\partial^2 f}{\partial x^{i_\mu} x^{j_\nu}} = \sum_{\nu\mu} v^{i_\mu} h^{j_\nu} \frac{\partial}{\partial x^{i_\mu}} (\beta_\nu D_j f(x)) =$$

$$\sum_{\mu} v^{i_\mu} h^j \frac{\partial}{\partial x^{i_\mu}} \frac{\partial}{\partial x^{j\phi}} f(x) = v^i h^j \frac{\partial}{\partial x^{j\phi}} \frac{\partial}{\partial x^{i\phi}} f(x) \equiv v^i h^j D_j D_i f(x)$$

Proposition 3.3. *Let* $\mathbb{Q}_p^{n,m} = (B_{p,k}^0)^n \times (B_{p,k}^1)^m$. *If* $f\colon \mathbb{Q}_p^{n,m} \to B_{p,k}$ *is* S^1 *and* C^ω *then* f *is* S^ω.

Any C^ω map f admits the expansion

$$f(x+h) = f(x) + Df(x)(h) + 1/2D^2 f(x)(h,h) + 1/6D^3 f(x)(h,h,h) + \ldots.$$

which can be written in the following required form by using S-differentiability

$$f(x+h) = f(x) + h^i D_i f(x) + 1/2 H^i h^j D_{ij}^2 f(x) + 1/6 H^i h^j h^k D_{ijk}^3 f(x) + \ldots.$$

There is a very interesting and surprising difference between real and p-adic superanalysis which is due to the fact that there is no Taylor formula for p-adic analysis with integral or Lagrangian form for the rest. More precisely, for every differentiable map $f\colon \mathbb{Q}_p \to \mathbb{Q}_p$ we can write

$$f(x+h) = f(x) + hf'(x) + o(h), \text{ where } \frac{o(h)}{h} \to 0, \text{ when } h \to 0$$

but we have no way to control the magnitude of the rest with integral or Lagrangian formulas. As a consequence, there are famous examples of maps $f\colon \mathbb{Q}_p \to \mathbb{Q}_p$ which are *not constant* and, however, have identically vanishing derivatives.

This peculiarity has a great consequence also in our study, since, while in the real case, any S^∞ map $f\colon \mathbb{R}^{n,m} \to \Lambda$ can be expanded with respect to its anticommuting variables

$$f(x,\theta) = f(x) + \sum_{k=1}^{m} \frac{1}{k!} \sum_{\alpha_1,\ldots,\alpha_k=1} \theta^{\alpha_1} \ldots \theta^{\alpha_k} D_{\alpha_k \ldots \alpha_1} f(x,0)$$

(this formula is called the superfield expansion of the function f), in the present p-adic case, this is no longer true.

Theorem 3.2 *If* $\Lambda = B_{p,0}$ *there exists* S^∞ *functions* $f\colon \mathbb{Q}_p^{0,m} \to \Lambda$ *which are not polynomial.*

In other words, while superfield expansion formula is true by definition for S^ω maps, superfield expansion formula is not true for every S^∞ function.

Proof. Let us consider $B_{p,0}$ superalgebra, and expand the odd variable θ as $\theta = \sum_\mu \theta^\mu \beta_\mu$. Denote with $O_1 = \{\theta \in B_{p,1} : \|\theta\|_{p,k} \le 1\}$ (the unitary clopen ball in $B_{p,1}$) and $Z_p = \{x \in \mathbb{Q}_p : |x|_p \le 1\}$ (the so-called p-adic integer ring) and consider the sequence of continuous maps $\{f_\nu(t)\}_{\nu \in M_\infty}$ where $f_\nu : Z_p \to \mathbb{Q}_p$ such that $\max_{t \in Z_p} |f_\nu(t)|_p \to 0$ whenever $\nu_1 + \ldots + \nu_n \to \infty$.

Let us define the map $f(\theta) = \sum_\mu f_\mu(\theta^\mu)\beta_\mu$. This map is well defined and continuous in the ball O_1 and valued in $B_{p,1}$. Assume now that the functions $f_\nu(t)$ are differentiable and such that $f'_\nu(t) = 0$; by using differentiability we get

$$f_\nu(\theta^\mu + h^\mu) = f_\nu(\theta^\mu) + o_\mu(h^\mu)$$

If we assume that $\max_\mu |o_\mu(h^\mu)|_p \to 0$ whenever $\|h\|_{p,k} \to 0$, the map f is S-differentiable and satisfies $Df = 0$, so it is S^∞ but it is not-necessarily constant: so it is not S^ω.

There is no problem to construct a function $f(\theta) \in S^\infty$ with non-zero superderivative which is not a polynome by using linear combinations of S-analytic functions and functions constructed as in the previous theorem. And there is also no problem to generalize the example of the previous theorem to an arbitrary superalgebra $B_{p,k}$.

Open problem *Is it possible to represent every S^∞ map $f(\theta)$ as a linear combination $f(\theta) = u(\theta)\theta + v(\theta)$, where u and v are S^∞ maps with zero superderivatives?*

In real and complex case we have only one S^∞-curve of the anti-commuting parameter θ. It is a line $y^i = a_i\theta + d_i, i = 1, ..., s$.

The situation is not so trivial in p-adic case. There is a large number of curves, $y^i = y^i(\theta)$ which are not straight lines. We can also consider non-polynomial S^∞-surfaces $Y^i = Y^i(\theta_1, ..., \theta_k), i = 1, ..., s$ in this case.

Proposition 3.4 *Let V, W, U be p-adic superspaces.*

Let $f: V \to W$, and $g: W \to U$ S^1 maps and $y = f(x)$. Then the map $g \circ f$ is S^1 and its differential is $D(g \circ f)(x) = Dg(y) \circ Df(x)$. If

$f: V \to W$ is $\mathcal{SD}^1$ and $Df(x): V \to W$ is a Λ-left-linear isomorphism, then there is an open $\Delta \subset V$ ($x \in \Delta$) and an open $\Delta' \subset W$ ($y = f(x) \in \Delta'$) such that $f|_\Delta$ is an homeomorphism of V in W. Moreover, there exists the inverse map g which is $\mathcal{SD}^1$, and satisfies

$$Dg(f(x)) = [Df(x)]^{-1}$$

Let $f: \mathbb{Q}_p^{m,n} \to \Lambda$ a $\mathcal{SD}^2$ map. If we denote with

$$f(x^1, \dots, x^m, \theta^1, \dots, \theta^n)$$

such a map, the second order derivatives satisfy the symmetric property:

$$\partial_i \partial_j f = \partial_j \partial_i f$$

for all *even* i, j. This is a consequence of Proposition 3.1 since any even super derivatives can be regarded as a Frechèt derivatives in the direction of the unit.

The situation is completely different a priori for odd derivatives. However, even if no superfield expansion is present in our theory, we can get the standard graded commutative property.

Let $f(\theta, \xi)$ a $\mathcal{SD}^2$ map from $\mathbb{Q}_p^{0,2}$ to Λ. Proposition 3.1 implies that

$$D_\xi \left(D_\theta f(\theta, \xi)(h)\right)(v) = D_\theta \left(D_\xi f(\xi, \theta)(v)\right)(h)$$

Now, by using the triviality of Λ_1 annihilator, we get

$$vh \frac{\partial^2 f}{\partial\theta\partial\xi} = hv \frac{\partial^2 f}{\partial\xi\partial\theta}$$

Proposition 3.5 *Let $f: \mathbb{Q}_p^{m,n} \to \Lambda$ a $\mathcal{SD}^2$ map. The second order derivatives are graded symmetric. If $x^A \in \mathbb{Q}_p^{n,m}$, $A = 1, \dots, n+m$ we have:*

$$\frac{F}{\partial_{x^A} \partial_{x^B}} = (-1)^{AB} \frac{F}{\partial_{x^B} \partial_{x^A}}$$

4. p-Adic Supermanifolds

Definition 4.1 *Let X a topological space; a (n,m)-dimensional p-adic S^k (or $\mathcal{SD}^k$) atlas on X is a triplet $\mathcal{A} = \{(U_\alpha, \psi_\alpha), \mathbb{Q}_p^{n,m})|\alpha \in J\}$ where U_α is an open covering on X; the map ψ_α is an homeomorphism of U_α in an open $V_\alpha \subset \mathbb{Q}_p^{n,m}$; (it is also called a coordinate map); "the transition functions" $\psi_\alpha \bullet \psi_\mu^{-1}: \psi_\mu(U_\mu \cap U_\alpha) \mapsto \psi_\mu(U_\alpha \cap U_\mu)$ are S^k (or $\mathcal{SD}^k$) for all α and $\mu \in J$.*

Definition 4.2. *Two S^k (or $\mathcal{SD}^k$) atlases $\mathcal{A}$ and $\mathcal{A}'$ are said to be compatible (or equivalent) if the functions $\psi'_\alpha \bullet \psi_\mu^{-1}$ defined on $\psi_\mu\left(U'_\mu \cap U_\alpha\right)$ are S^k (or $\mathcal{SD}^k$) for all U'_α and U_μ.*

In other words, if $\mathcal{A} \cup \mathcal{A}'$ is still an atlas. A (m,n)-dimensional p-adic S-differential structure for a manifold X is an equivalence class of atlases of X. A (m,n)-dimensional S^k (or $\mathcal{SD}^k$) p-adic supermanifold X is a manifold with a p adic S^k (or $\mathcal{SD}^k$) S-differential structure.

As in the theory of differential manifolds, two different, inequivalent atlases make the *same* manifold M into two different p S-differential manifolds.

Definition 4.3 *Two p-adic superdifferential structure S and S' (or $\mathcal{SD}$ and $\mathcal{SD}'$) for a manifold X are said to be equivalent if there exists a continuous isomorphism $\phi: X \to X'$ which is S^k (or $\mathcal{SD}^k$) when expressed on the atlases S and S' ($\mathcal{SD}$ and $\mathcal{SD}'$).*

Suppose that X and X' are p-adic S^k (or $\mathcal{SD}^k$) supermanifolds. By means of the definition of the p-adic differential structures, we say that a map $f: X \to X'$, is S^r (or $\mathcal{SD}^r$) , $r \leq k$, if its local representations $\psi_\mu \dot{f} \psi'^{-1}_\nu$ are S^r (or $\mathcal{SD}^r$) for all $\mu \in J, \nu \in J'$.

Definition 4.4 *Let X a p-adic supermanifold and let $x \in X$. Consider a chart $c = (U_\alpha, \psi_\alpha), \mathbb{Q}_p^{n,m}$ ($x \in U_\alpha$) and the couple (c, v) where $v \in \mathbb{Q}_p^{n,m}$. We shall say that two couples (c, v) and (c', v') ($x \in U'_\beta$), are equivalent in x if the superderivative of the map $\psi'_\beta \dot{\psi}_\alpha^{-1}$ transform v in v'.*

By using local coordinates x^i in U_α and x'^i in U'_α, this definition yields

$$v^{j'} = \sum_{i=1}^{n+m} v^i \frac{\partial x'^j}{\partial x^i}.$$

A tangent vector $u \in T_x X$ is a *class of equivalence* of couple (c, h).

This definition can be changed by assuming $v \in \Lambda^{n+m}$ and, in this case, we get the so called *total* tangent space $\tilde{T}_x X$.

Like in the ordinary theory of supermanifolds, if we consider the equivalence class of paths $\gamma: I \subset \mathbb{Q}_p \to X$ with *first order contact* in $x \in X$ we generate again the tangent space $\tilde{T}_x X$; however, no Λ or Λ_0 module basis exists: only in $\tilde{T}_x X$ we can consider the set of *derivations* $\frac{\partial}{\partial x^i}$ for $i = 1, ..., n+m$ which are a basis of $\tilde{T}_x X$.

In a similar way we define the *total cotangent space* $\tilde{T}^*_x X$ in $x \in X$ as the set of Λ-linear maps from $\tilde{T}_x X \to \Lambda$ which is clearly a Λ-module; every element of this space can be uniquely written in the form $f(v^1, \ldots, v^{n+m}) = \sum_{i=0}^{n+m} v^i a_i$ where $a_1 = f(1, 0. \ldots, 0)$ and so on.

Finally, we define the *cotangent space* $T^*_x X$ for $x \in X$ as the subspace of $\tilde{T}^*_x X$ of the maps f for which $a_i \in \mathbb{Q}_p^{n,m}$.

The action of the maps $f \in T^*_x$ on $T_x X$ produces only Λ_0 elements.

If one consider smooth maps $f: X \to \Lambda_0$, the differential $df = \sum_{i+1}^{n+m} dx^i \frac{\partial f}{\partial x^i}$ is an element of cotangent space. In this way we have the so called duality relations

$$\frac{\partial}{\partial x^i}(dx^j) = \delta_i^j$$

for $i, j = 1, ..., n+m$ and the standard action of the vectors of $T_x X$ is

$$\omega(Y) = Y(\omega) = Y^i \frac{\partial}{\partial x^i}(dx^j \omega_j) = Y^i \omega_i. \tag{4.1}$$

If we consider a map $f: X \to \Lambda$, the differential df is an element of the total cotangent space and equation (4.1) is true also for total spaces.

In a standard way we can define differential forms on X and differential maps.

Proposition 4.1 *Let X an $\mathcal{SD}^k$ p-adic manifold and $\xi^1, \ldots, \xi^{m+n}$ $\mathcal{SD}^k$ maps defined in an open U of X. The following conditions are equivalent:*

for all $x \in X$ there is an open $\Delta \subset U, x \in \Delta$ where $\xi^i|_\Delta$ are local coordinates;

$d\xi^i$ are a basis for $\tilde{T}_x^* X$.

We wish to stress the importance of the $\mathcal{SD}^k$ condition since it is necessary to use the inverse map theorem.

Higher order S^k differential forms can be constructed in standard way: given a function f, define $df = dx^i \frac{\partial f}{\partial x^i}$; if $\omega = dx^i \omega_i$ define $d\omega = -dx^i \wedge dx^j \frac{\partial \omega_i}{\partial x^j}$.

The properties $d^2 = 0$ is true in $\mathcal{SD}^k$.

Proposition 4.2 *Let X, Y two p-adic S^k manifolds and $\psi: X \mapsto Y$ an S^k map. Suppose that $\psi(x) = y$. We can define the push-forward map $\psi_*: T_x X \mapsto T_y Y$ and the pull-back map $\psi^*: T^*Y \mapsto T^*X$ as in Banach manifold theory.*

The map ψ^* is a S^k function, and both maps ψ^* and ψ_*, when calculated at a point, are described by GL_p matrices;

if $\psi: X \mapsto Y$ is a 1-1 and $\mathcal{SD}^k$, (it is an isomorphism), then ψ_* is a superdifferential map.

Now we consider balls in $\mathbb{Q}_p$. They are the set

$$\mathcal{U}_\alpha = \left\{ t \in \mathbb{Q}_p : |t - t_0|_p < \alpha \right\}$$

for every real α.

Definition 4.5. *Let $\mathcal{U}_\alpha$ a ball with $t_0 = 0$. A one-dimensional group of $\mathcal{SD}^k$ isomorphism of X to X (automorphisms) in $\mathcal{U}_\alpha$ is a map $\psi: t \in \mathcal{U}_\alpha \times X \to X$ satisfying*

$$\phi_t(\phi_s(x)) = \phi_{t+s}(x)$$

for every parameters $t, s \in \mathcal{U}_\alpha$.

The vector field $v(x)$ defined as $\frac{d\phi_t(x)}{dt}$ is called the field generated by the flow $\phi_t(x)$ in x.

The Lie derivative of a vector filed v of TX with respect to the motion ϕ_t (or, with respect to v), is defined as

$$\mathcal{L}_v y \equiv \lim_{t\to 0} \frac{-\phi_{t*}y + y}{t}$$

similarly, the Lie derivative of a covector ω of $\tilde{T}^*X$ is

$$\mathcal{L}_v \omega \equiv \lim_{t\to 0} \frac{\phi_{t*}\omega - \omega}{t}$$

Two examples. If $V = \mathbb{Q}_p^{m,n}$, $\phi_t(v^1, \ldots, v^{m,n}) = (v^1+t, \ldots, v^{m,n}+t)$ is a one-dimensional group of $\mathcal{SD}^k$ automorphisms of V for every t.

If $V = \mathbb{Q}_p^{m,n}$ and A is a *fixed* $gl_{p,k}(m,n)$ matrix, we consider the clopen set $\mathcal{U}_\alpha = \left\{t \in \mathbb{Q}_p : \ ||t|_p < \alpha|\alpha = p^{1/(1-p)}\|\varepsilon\,(A)\,\|^{-1}\right\}$. Then, for every $v \in V$, the map $\phi_t(v) = \exp tAv$ is a one-dimensional group of $\mathcal{SD}^k$ automorphisms of V for $t \in \mathcal{U}_\alpha$.

Theorem 4.1. *Let $v(x)$ a S^ω-tangent field of X; then there exists a local one parameter group of p-adic automorphisms tangent to v.*

Proof. One has to solve the differential equation

$$\frac{dx^i(t)}{dt} = v^i(x(t)), \ x^i(0) = a^i$$

Now a formal expansion of the required solution solve our problem from a local point of view.

We have no results concerning the possibility to prove the previous theorem for $\mathcal{SD}^k$ automorphisms.

Let X a $\mathcal{SD}^k$ manifold and let ψ_t an $\mathcal{SD}^k$ map. In this case the Lie derivative can be fruitfully calculated with the usual commutator operator:

$$\mathcal{L}_v y = [v, y] = vy - yv$$

Lie derivatives can be constructed also for differential forms; in this case we have the formula:

$$\mathcal{L}_v \omega = i_v d\omega + d\,(i_v \omega)$$

Our analysis will continue toward the direction of p-adic super Lie groups, see [19], [20] (we shall say that a p-adic super Lie group is a p-adic supermanifold with an S^ω group structure).

Bibliographical Remarks

We shall first consider the physical motivation of the mathematical theory exposed in this book.

The discussion of the *structure of space-time on fantastically small Planck scales* has a long history (see, for instance, [10]). In connection with this problem, an idea was repeatedly advanced, in different variants, of a *non-Archimedean structure of space-time on Planck scales* (see, for instance, [6], [103], [131]). The non-Archimedean physical models provoked great interest in connection with the string theory. The pioneer work [132] caused a splash of publications on *p-adic strings* [3]-[5], [25], [26], [28], [31]–[33], [89], [98], [124], [133]. However, it proved to be very difficult to interpret physical models of a level as high as a p-adic string. Therefore a thorough investigation began of simpler non-Archimedean models such as quantum mechanics and the field theory [51], [58], [59], [62], [64], [69], [72], [76], [112], [113], [122], [123], [127]–[130], [137]–[139].

We understand non-Archimedean models to be all physical models where some variables belong to the non-Archimedean number field K. Two practically independent approaches are possible here. With the first approach [112], [113], [122], [123], [127]–[130], [137]–[139], the real space–time $\mathbb{R}^4$ is replaced by K^4 (as a rule, $K = \mathbb{Q}_p$ is a field of p-adic numbers) but the wave functions assume ordinary complex values. With the second approach [51], [58], [59], [62], [64], [69], [72], [76], wave functions assume non-Archimedean values and the structure of space-time may be Archimedean or non-Archimedean.

This book was devoted to the mathematical problems of *non-Archimedean-valued physics*. The distribution theory is given on finite-dimensional and infinite-dimensional non-Archimedean spaces [64], [69] and serves as the basis for the development of the formalism of the *Gauss and Feynman non-Archimedean integration* [69], [72]. Its application to the quantum mechanics and the field theory are considered [64], [69], [72].

The p-adic probability and statistics appeared [67], [75] in connection with the problem of the statistic interpretation of the quantum theories with wave functions assuming values in quadratic extensions of fields of p-adic numbers. The ordinary probability theory based on Kolmogorov's axiomatics [94] proved to be inapplicable here. New, *non-Kolmogorov probability theories* appeared [71].

The p-adic white noise serves as a very important example of the Gaussian distribution on a finite-dimensional non-Archimedean space [60].

The theories of generalized functions with values in the fields of non-Archimedean numbers, considered in this paper [64], [72], [69] are analytic in character (the spaces of the test functions are different spaces of analytic functions). The use of the generalized functions in mathematical physics made it possible to solve a number of problems of quantum mechanics and the field theory with non-Archimedean-valued wave functions. It stands to reason that these are only the first steps in the use of generalized functions in the non-Archimedean-valued mathematical physics and an essential advance in the direction considered in [12]–[15], [120], [121] is possible here.

We shall consider in greater detail three papers [103], [6], [131] which are important landmarks on the way of development of the non-Archimedean physics. To all appearance, the idea of the non-Archimedean structure of space-time in quantum physics was considered for the first time in [103] in which it was stated that there existed incongruent space-time intervals which had to be measured with the aid of p-adic numbers. This paper did not find any response in physics and has been forgotten. The authors of [6] approached p-adic numbers in physics with the means of *quantum logic*. They proceeded from the *propositional algebra* connected with the quantum experiment and studied the problem of the algebraic structure of a linear space in which the representation of the corresponding lattice of the quantum logic was considered, namely, the problem of a number field for this representation. The question concerning the possibility of using the field of p-adic numbers was answered in the negative. Because of the negative answer, the investigations begun in [6] were terminated. The author of [131] returned to the problem of the non-Archimedean structure of space-time on Planck scales. Instead of giving general philosophical reasons [103], this author considered two specific physical models, the gravitation and the string theory. These models aroused considerable interest [3]–[5], [25], [26], [28], [31]–[33], [89], [98], [51], [58], [59], [62], [64], [69], [72], [76], [112], [113], [122], [123], [127]–[130], [137]–[139], served as the basis for the contemporary non-Archimedean physics.

As a comment to [6] it should be pointed out that the non-Archimedean-valued quantum physics must not be developed in the framework of the standard axiomatics (whereas [6] is, in essence, the first attempt of constructing such a physics). The axiomatics must be changed, in particular, the concept of Hilbert space must be extended. Such an alteration of axiomatics was carried out in [51], [58], [59], [62], [64], [69], [72], [76], where, in particular, a *non-Archimedean Hilbert space* was introduced and a new, non-Kolmogorov, probability theory was constructed in which probabilities may be p-adic numbers. It would be very interesting to complete investigations [6] according to the non-Archimedean representations of lattices in quantum logic with due account of the new results [51], [58], [59], [62], [64], [69], [72], [76],

The *p-adic probability and statistics* [60], [67], [71], [75], [83], [90]–[93] that appeared in the p-adic-valued physics may find applications in other natural sciences as well, they may be used to describe stochastic models in which relative frequencies oscillate in the field of real numbers but are stabilized in one of the fields of p-adic numbers. It has become possible to obtain new information, inaccessible to a "real" researcher, from a statistical sample. An idea was conceived of a *stratified structure of information* in a statistical sample. In addition to ordinary real information concentrated in real probabilities, a statistical sample may also contain information strata corresponding to p-adic probabilities.

The theory of non-Archimedean-valued Gauss and Feynman integrals [64], [72], [69] was greatly influenced by the investigations of the ordinary *infinite-dimensional analysis* [1], [11], [24], [37]–[45], [47], [48], [114]–[117], where the methods of generalized integration were used for measures on infinite-dimensional spaces as well as methods that were developed when the theory of integrals was constructed for distributions on superspace [46], [49], [50], [52], [55], [56].

It does not touch upon a wide range of problems connected with the models of mathematical physics over the field of non-standard numbers (see, for instance, [2]). I think that there are many similar ideas in p-adic and non-standard mathematical physics. But, at the moment, I cannot investigate this problem.

Open Problems

Non-archimedean valued quantum models generated a great number of new mathematical problems in the non-Archimedean analysis.

1. A mathematical theory of non-Archimedean Hilbert space.

   We have proposed only the definition. There are many open questions in connection with this definition. First of all, what are the restrictions to weight sequences $\lambda = (\lambda_n)$ and $\mu = (\mu_n)$ to generate the isomorphic Hilbert spaces $\mathcal{H}_\lambda$ and $\mathcal{H}_\mu$?

2. A spectral theory of operators in non-Archimedean Hilbert spaces. An analogue of the selfadjoint operator, unitary operator. Evolutional groups of unitary operators.

3. Unboundedness of the Gaussian distribution for $p = 2$.

4. A connection between Lebesgue and Volkenborn distributions.

5. Integration on infinite-dimensional locally convex spaces.

6. An analogue of Kolmogorov's theorem.

7. Non-Archimedean stochastic processes and stochastic differential equations.

8. The Feynman-Kac formular for differential and pseudo-differential equations.

Appendix

1. Expansion of Numbers in a Given Scale

As the natural numbers are the simplest among all numbers, we shall start from these object. Besides of the standard decimal expansion with respect to the scale of m, Chapter 1, (3.3), which corresponds to the measurement process with m-times increasing of the length unit 1, we can consider the following factorial expansion

$$n = a_1 1! + a_2 2! + \ldots + a_n n!\ , \tag{1.1}$$

where a_j are digits such that $0 \leq a_j \leq j$ for $j = 1, 2, \ldots, n$. This factorial expansion is not so well known as the standard, Ch. 1, (3.3), and we present the prove that every natural number n can be uniquely expanded into (1.1).

Proof. Suppose that n admits two representations in the form (1.1). So

$$n = a_1 1! + a_2 2! + \ldots + a_n n! = c_1 1! + c_2 2! + \ldots + c_n n!$$

Let k denote the greatest natural number such that $a_k \neq c_k$, i.e. $c_k > a_k$, say. Therefore $c_k - a_k \geq 1$, and

$$k! \leq c_k k! - a_k k! = a_1 \cdot 1! + \ldots + a_{k-1}(k-1)! - c_1 \cdot 1! - \ldots - c_{k-1}(k-1)!$$

$$\leq 1 \cdot 1! + 2 \cdot 2! + \ldots + (k-1)(k-1)! = k! - 1 < k!.$$

And it is impossible.

Now let s denote a natural number. Consider all the expansions of the form (1.1) with $n \leq s$. It is easy to calculate, the number of them is equal to $(1 + 1)(2 + 1)...(s + 1) = (s + 1)!$. Therefore the number of the expansions excluding zero expansion is $(s + 1)! - 1$. But different expansions (1.1) correspond to different numbers. On the other hand, any expansion (1.1) with $n \leq s$ generates a natural number $\leq 1 \cdot 1! + 2 \cdot 2! + ... + n \cdot n! = (n+1)! - 1 \leq (s+1)! - 1$. Thus every natural number $\leq (s+1)! - 1$ can be obtained as an expansion (1.1) where $n \leq s$.

The expansion (1.1) can be considered as the measurement algorithm where the unit 1 is increasing into n times on the nth step. For any natural number this algorithm has a finite number of steps. Now let us consider also a possibility of decreasing of 1 into n times on the nth step. If we assume that this decreasing algorithm can has an infinite number of steps, we get the factorial expressions

$$x = ... + \frac{a_{-k}}{(k+1)!} + ... + \frac{a_{-1}}{2!} + a_1 \cdot 1! + ... + a_n \cdot n!, \tag{1.2}$$

where a_j are integers and $0 \leq a_k, a_{-k} \leq k$. As we have seen infinite algorithms with the fixed coefficient of the decreasing of 1 correspond to real numbers. The same is valid for the algorithms (1.2).

There is a possibility to consider more general decreasing algorithms in the varying scale. Let $m_1, m_2, ...$ be an infinite sequence of natural numbers > 1, x a real number. Infinite sequences c_0, c_{-1}, c_{-2}, ... and $x_1, x_2, ...$ are defined as follows:

$$c_0 = [x], x_1 = x - c_0, c_{-1} = [m_1 x_1], x_2 = m_1 x_1 - c_{-1}, \tag{1.3}$$

$$c_{-2} = [m_2 x_2], ..., c_{-n} = [m_n x_n], ..., x_{n+1} = m_n x_n - c_{-n}, n = 1, 2, ...$$

It is clear that $0 \leq x_n < 1$ and $0 \leq c_{-n} \leq m_n - 1$. The algorithm (1.3) corresponds to the expansion of a real number x of the form

$$x = c_0 + \frac{c_{-1}}{m_1} + \frac{c_{-2}}{m_1 m_2} + ... + \frac{c_{-n}}{m_1 m_2 \dots m_n} + \frac{x_{n+1}}{m_1 m_2 \dots m_n}. \tag{1.4}$$

Since for $n = 1, 2, ...$ we have $m_n \geq 2$ and $0 \leq x_{n-1} < 1$, the last summand in (1.4) is non-negative and less than $1/2^n$, so it approaches zero when n approaches infinity. Thus we have the expansion of x in an infinite series:

$$x = c_0 + \frac{c_{-1}}{m_1} + \frac{c_{-2}}{m_1 m_2} + \ldots + \frac{c_{-n}}{m_1 m_2 \ldots m_n} + \ldots \tag{1.5}$$

If $m_n = n+1, n = 1, 2, \ldots$,it coincides with (1.2) and if $m_1 = m_2 = \ldots = m$, with (3.2) Chapter 1.

In the factorial case algorithm (1.3) has an interesting property of the characterization of rational numbers. It is possible to prove that if x is rational then (1.3) leads to a finite expansion in the form (1.2). Such result confirm our considerations of $\mathbb{Q}$ as a set "physical numbers" and other number systems as the only idealizations generated by our mind. If we use the algorithm (1.3) in the factorial case, we need only finite number of steps to measure any physical value.

However, any rational admits also another infinite representation in form (1.2). This follows from the following identity

$$x = \frac{c_{-k}}{(k+1)!} + \frac{c_{-(k-1)}}{k!} + \ldots + \frac{c_{-1}}{2!} + c_0$$

$$= \ldots + \frac{k+2}{(k+3)!} + \frac{k+1}{(k+2)!} + \frac{c_{-k} - 1}{(k+1)!} + \frac{c_{-(k-1)}}{k!} \ldots + \frac{c_{-1}}{2!} + c_0.$$

There are some other types expansions of real numbers. For example, the expansion with respect to the fixed scale where an arbitrary real number $g, g > 1$, is considered as the base instead of a natural number m. For example, we can consider the scale where the coefficient of decreasing of 1 is equal to $\sqrt{2}$. In such case we get expansions with respect to powers of $\sqrt{2}$. There are also negative scales and varying scales where the sequence $m_1, m_2, \ldots$ depends on the real number x.

But if our imagination can without great problem create a measurement process (1.2) which contains an infinite number of steps of decreasing of the unit of measurement 1, there is also no problem to study an analogous measurement process with an infinite number of steps of increasing of 1

$$x = \frac{a_{-k}}{(k+1)!} + \ldots + \frac{a_{-1}}{2!} + a_1 \cdot 1! + \ldots + a_n \cdot n! + \ldots \tag{1.6}$$

These measurement algorithms generate new algebraic structure on the basis of the field of rational numbers $\mathbb{Q}$. It would be possible to generalize main constructions of this book to this case. At the moment it is not easy to understand what is the main difference between p-adic (or m-adic) physical models and the models, describing by (1.5).

2. An Analogue of Newton's Method

Theorem 2.1. *Let $F(x), x \in Z_p$, be a polynomial with coefficients $F_i \in Z_p$. Let there exists such $\gamma \in Z_p$ that*

$$F(\gamma) = 0 (\text{mod } p^{2\delta+1}) \text{ and } F'(\gamma) = 0(\text{mod } p^{\delta}), F'(\gamma) \neq 0(\text{mod } p^{\delta+1}),$$

where δ is a natural number. Then there exists such p-adic integer α that

$$F(\alpha) = 0 \text{ and } \alpha = \gamma(\text{mod } p^{\delta+1}).$$

Proof. The conditions on the derivative of the polynomial can be rewritten in the form $F'(\gamma) = up^{\delta}$, where $|u|_p = 1$. We shall use the method which is very similar to the standard real Newton method.

Beginning from $\alpha_0 = \gamma$, we shall construct a sequence α_0, $\alpha 1$, ..., α_n, ..., with the aid of an iteration procedure:

$$\alpha_{n+1} = \alpha_n - \frac{F(\alpha_n)}{F'(\alpha_n)}. \tag{2.1}$$

Prove that all α_n are p-adic integers and

$$F(\alpha_n) = 0(\text{mod } p^{2\delta+1+n}), n \geq 0, \tag{2.2}$$

$$\alpha_n = \alpha_{n-1}(\text{mod } p^{\delta+n}), n \geq 1. \qquad (2.3).$$

We propose the proof of (2.2) and (2.3) with the aid of the induction with respect to n. Let (2.2) and (2.3) be valid for some $n \geq 0$(if $n = 0$ then we consider only (2.2)). As $\alpha_n = \alpha_0(\text{mod } p^{\delta+1})$, then $F'(\alpha_n) = F'(\alpha_0) = up^{\delta}$. Hence $F'(\alpha_n) = u_n p^{\delta}$, where $|u_n|_p = 1$.

Consequently, using (2.2), we get that α_{n+1} is a p-adic integer and $\alpha_{n+1} = \alpha_n (\text{mod } p^{\delta+n+1})$.

Further, expand the polynomial $F(x)$ with respect to powers of $x\alpha_n$

$$F(x) = F(\alpha_n) + F'(\alpha_n)(x - \alpha_n) + (x - \alpha_n)^2 G(x),$$

where $G(x)$ is a polynomial with p-adic integer coefficients. Set $x = \alpha_{n+1}$. Using (2.1), we get

$$F(\alpha_{n+1}) = \left(\frac{F(\alpha_n)}{F'(\alpha_n)}\right)^2 G(\alpha_n).$$

Hence $F(\alpha_{n+1}) = 0(\text{mod } p^{2\delta+2+2n})$. Thus, (2.2) and (2.3) valid for all n.

On the basis of (2.3), we have that the sequence $\{\alpha_n\}_{n=0}^{\infty}$ converges. Denote its limit by α. It is evident that $\alpha = \alpha_0 = \gamma(\text{mod } p^{\delta+1})$. Further, using (2.2), we get $\lim_{n\to\infty} F(\alpha_n) = 0$. On the other hand, $\lim_{n\to\infty} F(\alpha_n) = F(\alpha)$. Thus, $F(\alpha) = 0$.

Corollary 2.1. *Let $p(x)$ be a polynomial with p-adic integer coefficients and there exists $\gamma \in Z_p$ such that:*

$$F(\gamma) = 0(\text{mod } p), F'(\gamma) \neq 0(\text{mod } p).$$

Then there exists such $\alpha \in Z_p$ that

$$F(\alpha) = 0 and, \alpha = \gamma(\text{mod } p).$$

There is no problem to realize the square-root algorithm on a computer. Such a programme was written by my student V.V. Bezgin. It is a possibility to compute square roots for sufficiently large p with practically arbitrary number of p-adic digits. For example, $p = 11$ then $\sqrt{3} = 5268199439283491...$ We denote digits after 9 with the aid of English alphabet letters :$10 = A, 11 = B, 12 = C, 13 = D, ...$ and so on. For example, $p = 17$ then $\sqrt{2} = 6EE854EE72FF...$ or $\sqrt{13} = 93CG595D507B3F...$, or $\sqrt{15} = 7C6DF22A0F5...$ With the aid of this programme there is no problem to show quickly that square roots of 3, 5, 6, 7, 10, 11, 12, 14 do not exist in Q_{17}. For example, we can choose $Q_{17}(\sqrt{3}), Q_{17}(\sqrt{17})$ and $Q_{17}(\sqrt{51})$ as all quadratic extensions of Q_{17}.

3. Non-Existence of Differential Maps from $\mathbb{Q}_p$ to $\mathbb{R}$

We try to determine the existence of differential maps between Banach spaces over p-adic numbers field and real numbers field; our result will be a negative one: there are no maps of this kinds.

Now we give the definition of linear map between linear space constructed on different fields; this definition is a particular case of the more general one of functor between modules as given in the literature of category theory .

Definition 3.1. *Let B a Banach space over the field I and F a Banach space over the field K. A linear map A from $B \to F$ is a couple (A, g) where*

A is a homomorphism of the additive groups of the Banach spaces B and F, so $A(x + y) = A(x) + A(y)$ for all $x, y \in B$;

g is an homomorphism of the multiplicative groups of the fields I and K, so $g(\lambda\rho) = g(\lambda)g(\rho)$;

homomorphisms A and g are connected in the next way: $A(\lambda x) = g(\lambda)A(x)$ for $x \in B$ and $\lambda \in I, \lambda \neq 0$;

Now we define differential maps.

Definition 3.2. *A map $F: B \to F$ is Frechèt-differentiable in $x \in B$ if there is a continuous linear map $Df_x: B \to F$ such that for every $h \in B$ we have*

$$f(x + h) = f(x) + Df_x(h) + o(h)$$

where $\frac{\|o(h)\|}{\|h\|} \to 0$ whenever $\|h\| \to 0$.

Now we state our main theorem.

Theorem 3.1. *The only Frechèt-differentiable maps from $\mathbb{Q}_p$ to $\mathbb{R}$ have zero derivative; the same is true for maps form $\mathbb{Q}_p \to Q_r$ $p \neq r$.*

Proof. Denote with A the differential Df_x and let g be the associated homomorphism from the multiplicative group of $\mathbb{Q}_p$ to $\mathbb{R}$. Then we get $A(\alpha + \beta) = A((\alpha + \beta)1) = g(\alpha + \beta)A(1)$ and $A(\alpha) + A(\beta) = [g(\alpha) + g(\beta)]A(1)$ Since $A(1) \neq 0$ then

$$g(\alpha + \beta) = g(\alpha) + g(\beta).$$

Using this equality we get that $g(n) = n$ for every natural number n. In the same way equation $g(m/m) = g(1) = 1$ implies $g(x) = x$ for every positive rational number x. Consider now the sequence of positive rational numbers $x_n = 1+\frac{2+p^n}{1+p^{2n}}$ for $n = 1, 2, 3...$ This sequence converges to 3 in $\mathbb{Q}_p$ and to 1 in $\mathbb{R}$. Now,

$$g\left(\lim_{n\to\infty} x_n\right) = g(3) = 3$$

while

$$\left(\lim_{n\to\infty} g(x_n)\right) = \lim_{n\to\infty} x_n = 1$$

In the same way, we can show that the unique differentiable maps from $\mathbb{Q}_p$ to $\mathbb{C}$ have zero derivative.

The last statement can be proved by considering the sequence $x_n = p^n$.

Bibliography

1. Albeverio S. and Hoegh-Krohn P., *Mathematical Theory of Feynman Path Integrals*, Lecture Notes in Math. Springer, Berlin, 523, (1978).

2. Albeverio S., Fenstad I., Hoegh-Krohn P., and Lindström T ., *Nonstandard Method in the Stochastic Analysis and Mathematical Physics*, Academic Press, London, ... (1990).

3. Aref'eva I.Ya., Dragovic B., and Volovich I. V., On the p-adic summability of the anharmonic oscillator. *Phys. Lett. B.* , **200**, 512-514 (1988).

4. Aref'eva I.Ya., Dragovic B., and Volovich I. V., Open and closed p-adic strings and quadratic extensions of number fields. *Phys. Lett. B* ., **200**, 512-516 (1988).

5. Aref'eva I. Ya., Dragovich B., Frampton P. H., Volovich I. V. The wave function of the Universe and p-adic gravity. *Int. J. of Modern Phys., A*, **6**, No 24, 4341-4358 (1991).

6. Beltrametti E. and Cassinelli G., Quantum mechanics and p-adic numbers.*Found. Phys.*, **2**, 1-7, (1972).

7. Berezansky Yu. M., *Self-adjoint Operators in the Space of Functions of an Infinite Number of Variables*, Amer. Math. Soc., Providence, R.I. (1986).

8. Berezin F. A., The functional integral with respect to paths in a phase space. *Uspekhi Fiz. Nauk*, **132**, 497-548 (1980).

9. Bezgin V. V., Khrennikov A. Yu., Endo M., Uoko M. Statistical biological model with the p-adic statistical stabilization.*Dokl. Acad. Nauk.*, **334**, No 1, 5-8 (1994).

10. Blokhintsev D. I., *Space and Time in the Microcosm*, Nauka, Moscow (1982).

11. Bogachev V . I. and Smolaynov O. G., Analytic properties of infinite-dimensional distributions. *Uspekhi Mat. Nauk*, **45**, No. 2, 3-84 (1990).

12. Bogolyubov N. N. and Shirkov D. V., *Introduction to the Theory of Quantized Fields*, Nauka, Moscow (1984).

13. Bogolyubov N. N., Logunov A. A., and Todorov I. T., *Fundamentals of the Axiomatic Approach to the Quantum Field Theory*, Nauka, Moscow (1969).

14. Bogolyubov N. N., Bogolyubov N. N., Jr., *Introduction to the Quantum Statistical Mechanics*, Nauka, Moscow (1984).

15. Bogolyubov N. N., Medvedev B. N., and Polivanov M. K., *Problems of the Theory of Dispersion Relations*, Fizmatgiz, Moscow (1958).

16. Borel E., *Probability and Certainty* . Nauka, Moscow (1970).

17. Borevich Z. I. and Shafarevich I. R., *The Number Theory*, Academic Press, New-York, 1966.

18. Cianci R. and Khrennikov A. Yu., p-adic superanalysis, 1. Infinitely generated Banach superalgebras and algebraic groups. *J. Math. Phys.*, **34**, No. 5, 1990-1995, (1993).

19. Cianci R. and Khrennikov A. Yu., p-adic superanalysis, 2. Supermanifolds and differential operators, *J. Math. Phys.*, **34**, No. 5, 1995-1999, (1993).

20. Cianci R. and Khrennikov A. Yu., p-adic superanalysis, 3. Super-Lie groups, *J. Math. Phys.*, **35**, No.3(1994).

21. Cianci R. and Khrennikov A. Yu. , Can p-adic numbers be useful to regularize divergent expectation values of quantum observables ? Preprint, Depart. Math., Univ. Genova, November (1993). To be published in *Int. J. of Theor. Phys.*, 1994.

22. Cianci R. and Khrennikov A.Yu. Non-archimedean superfields. *Izvestia VUZOV, ser. Fizika*, No 11, 35-38 (1993).

23. Cramer H., *Mathematical theory of statistics.* Univ. Press, Princeton (1949).

24. Daletsky Yu. L. and Fomin S. V., *Measures and Differential Equations in Infinite-Dimensional Spaces*, Nauka, Moscow (1983).

25. Dragovic B., On signature change in p-adic space-time. *Mod. Phys. Lett.*, **6**, No. 25, 2301-2307, (1991).

26. Dragovic B., p-adic perturbation series and adelic summability. *Phys. Lett. B* , **256**, No. 3/4, 392-396, (1991).

27. Escassut A., T-filters, ensembles analytiques et transformation de Fourier p-adique. *Ann. Inst. Fourier*, **25**, No. 2, 45-80 (1975).

28. Frampton P. H. and Okada Y., p-adic string N-point function. *Phys. Rev. Lett. B* , **60**, 484-486 (1988).

29. Frechet M., *Recherches theoriques modernes sur la theorie des probabilites. Univ. Press.*, Paris (1937-1938).

30. Fresnel J. and De Mathan B., Sur la transformation de Fourier p-adique, *C. R. Acad. Sci. Paris*, **277**, 711-714 (1973).

31. Freund P. G. O. and Olson M., Non-Archimedean strings. *Phys. Lett. B*, **199**, 186-190 (1987).

32. Freund P. G. O. and Witten E., Adelic string amplitudes.*Phys. Lett. B* , **199**, 191-195 (1987).

33. Gervais J. L., *p*-adic analyticity and Virasoro algebras for conformal theories in more than two dimensions.*Phys. Lett. B* , **201**, 306-310 (1988).

34. Hida T., *Brownian Motion*, Nauka, Moscow (1987).

35. Hida T. and Streit L., On quantum theory in terms of white noise analysis. *Nagoya Math. J.*, **68**, 21-34 (1977).

36. Hida T. and Streit L., Generalized Brownian functionals and the Feynman integral.*Stoch. Process Appl.*, **16**, 55-69 (1983).

37. Khrennikov A. Yu., Equations with infinite-dimensional pseudodifferential operators.*Dokl. Akad. Nauk SSSR*, **267**, No. 6, 1313-1318 (1982).

38. Khrennikov A. Yu., Some applications of Keller's convergence structure to the nonlinear functional analysis. *Abh. Acad. Wiss., DDR*, 1984, 2, 112-118 (1984).

39. Khrennikov A. Yu., An existence theorem for the solution of the infinite- dimensional Schrodinger equation with qudratic potential. *Uspehi Matem. Nauk*, **39**, No 1, 163-164 (1984).

40. Khrennikov A. Yu., Fundamental solutions of evolutional pseudodifferential equations .*Diff. Equations*, **21**, No 2, 346-348 (1985).

41. Khrennikov A. Yu., Differential equations in locally convex spaces and evolutional pseudo-differential equations.*Diff. Equations*, **22**, No 9, 1596-1602 (1986).

42. Khrennikov A. Yu., On a theory of generalized measures on a Hilbert space. *Vestn. Mosk. Univ., ser. Matem.*, No. 2, 81-83 (1985).

43. Khrennikov A. Yu., Feynman's measure in a phase space and the symbols of infinite-dimensional pseudo-differential operators. *Mat. Zametki*, **37**, No. 5, 734-742 (1985).

44. Khrennikov A. Yu., Integration with respect to generalized measures on topological linear spaces. *Proc. of Moscow Math. Soc.*, **49**, 113-129 (1986).

45. Khrennikov A. Yu., Secondary quantization and pseudo-differential operators. *Theor. and Math. Phys.*, **66**, No. 3, 339-349 (1986).

46. Khrennikov A. Yu., Superanalysis: generalized functions and pseudo-differential operators. *Theor. and Math. Phys.*, **73**, No. 3, 420-429 (1987).

47. Khrennikov A. Yu., Infinite-dimensional pseudo-differential operators. *Izv. Akad. Nauk SSSR, Ser. Mat.* , **51**, No. 6, 46-68 (1987).

48. A. Yu. Khrennikov, Feynman's measures on locally convex spaces. *Sibir. Mat. J.*, **29**, No. 4, 180-189 (1988).

49. Khrennikov A. Yu., Pseudodifferential equations in the functional superanalysis, 1. Method of Fourier transformations. *Differ. Equations*, **24**, No. 12, 2144-2154 (1988).

50. Khrennikov A. Yu., Pseudodifferential equations in the functional superanalysis, 2. The Feynman-Kac formula. *Differ. Equations*, **25**, No. 2 (1989).

51. Khrennikov A. Yu., Quantization on the non-Archimedean superspace. *Proc. of Int. Conf. Actual Problems of Complex Analysis*, Tashkent, 129 (1989).

52. Khrennikov A. Yu., The principle of correspondence in the quantum field theory and in the theory of the relativistic boson string. *Mat. Sbornik.*, **180**, No. 6, 763-786 (1989).

53. Khrennikov A. Yu., The quantization of a boson string field and infinite-dimensional pseudodifferential operators. Fixed gauge. *Teor. and Math. Phys.*, **80**, No. 2, 226-238 (1989).

54. Khrennikov A. Yu., The quantization of a boson string field and Klein-Gordon nonlinear equation. *Izv. VUZOV. Phys. series*, **33**, No. 1, 56-60 (1990).

55. Khrennikov A. Yu., Equations on a superspace. *Izv. Akad. Nauk SSSR. Ser. Mat.*, **54**, No. 3, 576-606 (1990).

56. Khrennikov A. Yu., Functional superanalysis. *Uspekhi Mat. Nauk*, **43**, No. 2, 72-114 (1988).

57. Khrennikov A. Yu., Pseudodifferential operators on non-Archimedean spaces. *Differ. Equations*, **26**, No. 6, 1044-1053 (1990).

58. Khrennikov A. Yu., Representations of Schrödinger and Bargmann-Fock in non-Archimedean quantum mechanics. *Dokl. Akad. Nauk SSSR, ser. Fiz.*, **313**, No. 2, 325-329 (1990).

59. Khrennikov A. Yu., Representation of the second quantization over non-Archimedean number fields. *Dokl. Akad. Nauk SSSR, ser. Fiz.* , **314**, No. 6, 1380-1384 (1990).

60. Khrennikov A. Yu., Non-Archimedean White Noise. *Proc. Int. Conf. on Gaussian Random Fields*, Nagoya, 127 (1990).

61. Khrennikov A. Yu., Pseudotopological commutative superalgebras with nilpotent ghosts. *Math. Notes*, **48**, 114-122 (1990).

62. Khrennikov A. Yu., Quantum mechanics over the extensions of the Galois number fields. *Dokl. Akad. Nauk SSSR, ser. Fiz.*, **315**, No. 4, 860-864 (1990).

63. Khrennikov A. Yu., Trotter's formula for heat conduction equations and for Schrödinger's equation on non-Archimedean superspace. *Sibir Mat. J.*, **32**, No. 5, 155-165 (1991).

64. Khrennikov A. Yu., Mathematical methods of the non-Archimedean physics. *Uspekhi Mat. Nauk*, **45**, No. 4, 79-110 (1990).

65. Khrennikov A. Yu., Real-non-Archimedean structure of space-time. *Teor. and Matem. Phys.*, **86**, No. 2, 177-190 (1991).

66. Khrennikov A. Yu., Generalized functions on a non-Archimedean superspace. *Izv. Akad. Nauk SSSR, Ser. Mat.*, **55**, No. 6, 1257-1286 (1991).

67. Khrennikov A. Yu., The p-adic probability and statistics. *Dokl. Akad. Nauk SSSR, ser. Matem.*, **322**, No. 6, 1075-1079 (1992).

68. Khrennikov A. Yu., Infinite-dimensional Liouville equation. *Mat. Sbornik.*, **183**, No. 1, 20-44 (1992).

69. Khrennikov A. Yu., Generalized functions and Gaussian functional integrals with respect to non-Archimedean functional spaces. *Izv. Akad. Nauk SSSR, Ser. Mat.*, **55**, No. 4, 780-814 (1991).

70. Khrennikov A. Yu., Endo M., The unboundedness of the p-adic Gaussian distribution. *Izv. Akad. Nauk SSSR, Ser. Mat.* , **56**, No. 4, 456-476 (1992).

71. Khrennikov A. Yu., Axiomatics of the p-adic theory of probabilities. *Dokl. Acad. Nauk SSSR, sr. Matem.*, **326**, No. 5, 1075-1079 (1992).

72. Khrennikov A. Yu., p-adic quantum mechanics with p-adic valued functions. *J. Math. Phys.*, **32**, No. 4, 932-937 (1991).

73. Khrennikov A. Yu., Analysis on p-adic superspace, 1. Generalized functions: Gaussian distribution. *J. Math. Phys.*, **33**, No. 5, 1636-1642 (1992).

74. Khrennikov A. Yu., Analysis on p-adic superspace, 2. Differential equations. *J. Math. Phys.*, **33**, No. 5. 1643-1647 (1992).

75. Khrennikov A. Yu., *Non-Archimedean Theory of Probability: Frequency and Axiomatic Theories*, Preprint, Depart. Math., Univ. Genova. No. 209 (1992).

76. Khrennikov A. Yu., *p-adic valued quantization.* Preprint, Depart. Math., Univ. Genova, No 215 (1992). To be published in *Int. J. of Theor.Phys.*, 1993.

77. Khrennikov A. Yu., On a theory of the infinite-dimensional superspace : reflexive Banach supermodules. *Math. Sbornik*, **183** 75-98 (1992).

78. Khrennikov A. Yu., Formular of integration by parts for Feynman and Gaussian distributions on the superspace. *Izvestia Vuzov, ser.Math.*, **4**, 126-133 (1990).

79. Khrennikov A. Yu., Hilbert superspace.*Dokl. Acad. Nauk SSSR, ser. Matem.*, **321**, 298-301 (1991).

80. Khrennikov A. Yu., Noncommutative differential calculus and projective tensor products of noncommutative Banach algebras. *Dokl.Acad.Nauk SSSR, ser. Matem.*, **321**, 722-726 (1991).

81. Khrennikov A. Yu. , Generalized functions on infinite-dimensional spaces of sources. *Dokl. Acad. Nauk, ser. Fiz.* , **322** , 692-696 (1992).

82. Khrennikov A. Yu., Central limit theorem for quazi-Gaussian distribution on an infinite-dimensional superspace. *Theor. of Prob. and Appl.*, **35** , 599-602 (1990).

83. Khrennikov A. Yu., The theory of probability and the theory of numbers. *Com. Math. Univ. Sancti Pauli (Rikkyo Univ.)*, **41**, No 2, 135-140 (1992).

84. Khrennikov A. Yu., Process of Feynman-Wiener on the superspace. *Theor. of Probab. and Appl.*, **38**, No 3 (1993).

85. Khrennikov A. Yu., Fundamental solutions over the field of p-adic numbers. *Algebra and Analysis(Leningrad Math.J.)*, **4**, No 3, 248-266 (1992).

86. Khrennikov A. Yu., Formula of Feynman-Katz on the phase superspace, 1. *Diff.Equations*, **28**, No 8, 1434-1443 (1992).

87. Khrennikov A. Yu., Formula of Feynman-Katz on the phase superspace, 2. *Diff. Equations*, **28**, No 9, 1599-1607 (1992).

88. Khrennikov A. Yu., *Generalized functions with non-Archimedean values and their applications to quantum mechanics and field theory.* Analysis-3, Modern Mathematics and its Applications, VINITI, Moscow (1993) .

89. Khrennikov A. Yu., Gauge connection between real and p-adic space-time, *Proc.of Marcel Grossmann Meeting on General Relativity*, Kyoto, 1991, (Ed. H. Sato, T. Nakamura). World scientific publ., Singapore, 498-500 (1992).

90. Khrennikov A. Yu. , p-adic theory of probability and its applications. A principle of the statistical stabilization of frequencies. *Teor. and Matem. Fiz.*, **97**, No 3, 348-363 (1993).

91. Khrennikov A. Yu. , Statistical interpretation of p-adic valued quantum field theory .*Dokl. Acad. Nauk., ser. Fiz.*, **328**, No 1, 46- 50 (1993).

92. Khrennikov A. Yu. , p-adic statistical models. *Dokl. Akad. Nauk., ser.Matem.*, **330**, No 3, 300-304 (1993).

93. Khrennikov A. Yu. , Discrete Q_p-Valued probabilities. *Dokl. Acad. Nauk, ser. Matem.*, **333**, No 2, 162-164 (1993).

94. Kolmogorov A. N., *Fundamental Concepts of the Probability Theory*, Nauka, Moscow (1974).

95. Kolmolgorov A. N. Logical basis for information theory and probability theory . *IEEE Trans.*, **IT-14**, 662-664 (1968).

96. Konopleva N. and Popov. N., *Gauge Fields*, Atomizdat, Moscow (1980).

97. Mahler K., *p-adic Numbers and their Functions.* Cambridge tracts in math., 76. Cambridge Univ. Press, Cambridge (1980).

98. Makhaldiani N., *Number Fields Dynamics and the Problem of Compactification of a Space in the Unique Theories of Fields and Strings*, Preprint OIYaI, Dubna (1989).

99. Mises R., *Probability, Statistics and Truth*, Macmillan, London (1957).

100. Mises R., *The mathematical theory of probability and statistics.* Academic, London, ...(1964).

101. Monna A., *Analyse non-Archimedienne*, Springer, New-York, (1970).

102. Monna A. and Springer T., Integration non-Archimedianne, 1, 2. *Indag. Math.*, **25**, 634-653 (1963).

103. Monna A. and Blij F., Models of space and time in elementary physics. *J. Math. Anal. and Appl.*, **22**, 537-545 (1968).

104. Pontryagin L. S., *Topological Groups*, Princeton University Press, Princeton (1958).

105. Popov V. N., *Functional Integrals in the Quantum Field Theory and Statistical Physics*, Atomizdat, Moscow (1976).

106. Schikhov W., *Ultrametric calculus*, Cambridge Univ. Press, Cambridge (1984)

107. Schikhov W. and Y. Morita, Duality of projective limit spaces and inductive limit spaces over a nonspherically complete non-Archimedean field. *Tohoku Math. J.*, **38**, No. 3, 387-397 (1986).

108. Schikhov W., *p*-adic nonconvex compactoids. *Proc. Konik. Ned. Acad.*, **92**, No. 3, 339-342 (1989).

109. Schikhov W., Topological stability of p-adic compactoids under continuous injections. *Reports of Catholic Uni. Nijmegen.*, No. 8644 (1986).

110. Schikhov W., *Non-Archimedean Harmonic Analysis*, Nijmegen: Catholic Univ. Press. (1967).

111. Slavnov A. A. and Faddeev L. D., *Gauge Fields*, Benjiamin/Cummings, Reading, Mass. (1980).

112. Smirnov V.A., Calculation of general p-adic Feynman amplitude, it Commun. Math. Phys., **149**, 623-633 (1992).

113. Smirnov V. A., Renormalization in p-adic quantum field theory, *Mod. Phys. Lett.*, **6**, 1421-1427 (1991).

114. Smolaynov O. G. and Shavgulidze E. T., *Functional Integrals*, Moscow University Press, Moscow (1990).

115. Smolaynov O. G., Infinite-dimensional pseudodifferential operators and Schrödinger quantization. *Dokl. Akad. Nauk SSSR*, **263**, No. 3, 558-561 (1982).

116. Smolaynov O. G. and Khrennikov A. Yu., The Central limit theorem for generalized measures on an infinite-dimensional space. *Dokl. Akad. Nauk SSSR*, **281**, No. 2, 279-283 (1985).

117. Smolaynov O. G. and Khrennikov A. Yu., The algebra of infinite-dimensional pseudodifferential operators. *Dokl. Akad. Nauk SSSR*, **292**, No. 6, 1310-1313 (1987).

118. Smolaynov O. G. and Fomin S., Measures on topological linear spaces. *Uspekhi Mat. Nauk*, **31**, 3-56 (1976).

119. Tornier E., *Wahrtscheinlichkeitsrechnunug und allgemeine Integrationstheorie.*, Univ. Press, Leipzing (1936).

120. Vladimirov V. S., *Generalized Functions in Mathematical Physics*, Mir, Moscow (1979).

121. Vladimirov V. S., *The Methods of the Theory of Functions of Many Complex Variables*, Nauka, Moscow (1964).

122. Vladimirov V. S., Generalized functions over the field of p-adic numbers. *Uspekhi Mat. Nauk*, **43**, No. 5, 17-53 (1988).

123. Vladimirov V. S., On the spectrum of certain pseudodifferential operators over the field of p-adic numbers. *Algebra and Analysis*, No. 2, 107-124 (1990).

124. Vladimirov V. S. , On the Freund-Witten adelic formula for Veneziano amplitudes, *Lett. Math. Phys.*, **27**, 123-131 (1993).

125. Vladimirov V. S. and Volovich I. V., Superanalysis, 1. Differential Calculus. *Teor. and Matem. Fiz.*, **59**, No. 1, 3-27 (1984).

126. Vladimirov V. S. . and Volovich I. V., Superanalysis, 2. Integral Calculus. *Teor. and Matem. Fiz.*, **60**, No. 2, 169-198 (1984).

127. Vladimirov V. S. and Volovich I., The p-adic quantum mechanics. *Dokl. Akad. Nauk SSSR, ser. Fiz.*, **302**, No. 2, 320-322 (1988).

128. Vladimirov V. S. and Volovich I. V., p-adic quantum mechanics. *Commun. Math. Phys.*, **123**, 659-676 (1989).

129. Vladimirov V. S., Volovich I. V., and Zelenov E. I., *p-adic Numbers in Mathematical Physics*, World Scientific Publ., Singapoure (1993)

130. Vladimirov V. S., Volovich I. V., and Zelenov E. I., The spectral theory in the p-adic quantum mechanics. *Izv. Akad. Nauk SSSR, Ser. Mat.*, **54**, No. 2, 275-302 (1990).

131. Volovich I. V., *Number Theory as the Ultimate Physical Theory*, Preprint, CERN, Geneva. TH. 4781/87 (1987).

132. Volovich I. V., p-adic string. *Class. Quant. Gra.*, **4**, 83-87 (1987).

133. Volovich I. V., Harmonic analysis and p-adic strings. *Lett. Math. Phys.*, **16**, 61-64 (1988).

134. Uglanov A. V., On one construction of the Feynman integral, **243**, No. 6, 1406-1409 (1978).

135. Woodcock C. F., Fourier analysis for p-adic Lipschitz functions. *J. London Math. Soc.*, **7**, 681-683 (1974).

136. Woodcock C. F., Convolutions on the ring of p-adic integers. *J. London Math. Soc.*, **20**, 101-108 (1979).

137. Zelenov E. I., The p-adic quantum mechanics for $p = 2$. *Teor. Matem. Fiz.*, **54**, No. 2, 253-263 (1989).

138. Zelenov E. I., p-adic path integrals. *J. Math. Phys.*, **32**, 247-251 (1991).

139. Zelenov E. I., p-adic Heisenberg Group, Maslov Index. *Commun. Math. Phys.*, **155**, 489-502 (1993).

140. Zvonkin A. K. and Levin L. A. A complexity of finite objects and justification of the notions of information and randomness on the basis of algorithms. *Uspehi Matem. Nauk*, **25**, No 6, 85-127 (1970).

INDEX

MIX
Papier aus verantwortungsvollen Quellen
Paper from responsible sources
FSC® C105338

If you have any concerns about our products,
you can contact us on
ProductSafety@springernature.com

In case Publisher is established outside the EU,
the EU authorized representative is:
Springer Nature Customer Service Center GmbH
Europaplatz 3, 69115 Heidelberg, Germany

Printed by Libri Plureos GmbH
in Hamburg, Germany